Globalisation and Services-driven Economic Growth

Following drastic shifts in the spatial organisation of industrial goods production, increasingly fierce competition now forces firms also to look critically at how to organise the production of services. While digitisation and advances in information and communication technologies have enabled firms to unbundle service production processes, the increased global availability of skilled labour allows for the relocation of ever more of these processes around the world. As a result, a new geography of services production takes shape: a geography that is defined by new interregional and international divisions of labour and held together by increasingly complex global services production networks. This book examines how the reorganisation of services production alters relations between and generates different sets of challenges and opportunities for economic development in the Global North and the Global South.

Drawing from 11 case studies probing various aspects of services production in different parts of the world, the book brings out the remarkable heterogeneity and transformative capacities of services. It successively shows how global trade in services creates new interdependencies between services producing and services consuming regions; reveals how services help to mitigate the impact of and contribute to recovery from economic crises in the Global North; and demonstrates how services offshoring fosters economic development and service sector-driven modernisation processes in the Global South. The book's openness to the heterogeneous and dynamic nature of services production enlarges our understanding of which particular services in which spatiotemporal context have the capacity to generate good jobs, contribute to productivity and drive economic growth.

The book stands out from other books in the field in that it combines perspectives on services-driven transformations from both the Global North and the Global South and looks into the role of various services segments. Based on pioneering empirical research and original data it offers a timely contribution to this growing debate. The book provides valuable insights for students, scholars and professionals interested in services, services offshoring, services-driven economic growth, and socioeconomic transformations in the Global North and South.

Niels Beerepoot is Assistant Professor in Economic Geography at the University of Amsterdam, the Netherlands.

Bart Lambregts is post-doc researcher at the Amsterdam Institute for Social Science Research (AISSR) of the University of Amsterdam, the Netherlands. He currently lectures on urban planning at the Faculty of Architecture of Kasetsart University in Bangkok, Thailand.

Jana Kleibert is post-doc researcher at the Leibniz Institute for Research on Society and Space (IRS) in Erkner, Germany, and is honorary research fellow at the School of Environment, Education and Development at the University of Manchester, UK.

The Dynamics of Economic Space

Series Editor: Christine Tamásy, University of Vechta, Germany

The IGU Commission on 'The Dynamics of Economic Space' aims to play a leading international role in the development, promulgation and dissemination of new ideas in economic geography. It has as its goal the development of a strong analytical perspective on the processes, problems and policies associated with the dynamics of local and regional economies as they are incorporated into the globalizing world economy. In recognition of the increasing complexity of the world economy, the Commission's interests include: industrial production; business, professional and financial services, and the broader service economy including e-business; corporations, corporate power, enterprise and entrepreneurship; the changing world of work and intensifying economic interconnectedness.

Books in this series

Globalisation and Services-driven Economic Growth
Edited by Niels Beerepoot, Bart Lambregts and Jana Kleibert

Understanding Innovation in Emerging Economic Spaces
Edited by Grzegorz Micek

Economic Development in Rural Areas
Edited by Peter Dannenberg and Elmar Kulke

Regional Resilience, Economy and Society
Edited by Christine Tamásy and Javier Revilla Diez

Local Food Systems in Old Industrial Regions
Edited by Neil Reid, Jay D. Gatrell, and Paula S. Ross

Industrial Transition
Edited by Martina Fromhold-Eisebith and Martina Fuchs

Missing Links in Labour Geography
Edited by Ann Cecilie Bergene, Sylvi B. Endresen and Hege Merete Knutsen

Globalising Worlds and New Economic Configurations
Edited by Christine Tamásy and Mike Taylor

Globalisation and Services-driven Economic Growth

Perspectives from the Global North and South

Edited by Niels Beerepoot, Bart Lambregts and Jana Kleibert

Routledge
Taylor & Francis Group
LONDON AND NEW YORK

First published 2017 by Routledge

2 Park Square, Milton Park, Abingdon, Oxfordshire OX14 4RN
52 Vanderbilt Avenue, New York, NY 10017

Routledge is an imprint of the Taylor & Francis Group, an informa business

First issued in paperback 2020

British Library Cataloguing in Publication Data
A catalogue record for this book is available from the British Library

Library of Congress Cataloging in Publication Data
Names: Beerepoot, Niels, editor. | Lambregts, Bart, editor. | Kleibert, Jana, editor.
Title: Globalisation and Services-driven Economic Growth : perspectives from the global North and South / edited by Niels Beerepoot, Bart Lambregts and Jana Kleibert.
Description: Abingdon, Oxon ; New York, NY : Routledge, 2017. | Includes index.
Identifiers: LCCN 2016010437| ISBN 9781472470133 (hardback) | ISBN 9781315585055 (e-book)
Subjects: LCSH: Offshore outsourcing. | Service industries. | International trade.
Classification: LCC HD2365 .G62 2017 | DDC 382/.45—dc23
LC record available at https://lccn.loc.gov/2016010437

ISBN: 978-1-4724-7013-3 (hbk)
ISBN: 978-0-367-66814-3 (pbk)

Typeset in Times New Roman
by FiSH Books Ltd, Enfield

Contents

Figures

Tables

Abbreviations

ADB	Asian Development Bank
AHT	Average Handling Time
BLI	Brand Luxury Index
BPeSA	Business Processing enabling South Africa
BPO	Business Process Outsourcing
BRIIAT	Brazil, Russia, India, Indonesia, Australia, Turkey
CHED	Commission on Higher Education
CSAT	Customer Satisfaction Ratings
DCH	Direct Channel Holdings
EG	Ellison and Glaeser
EMEA	Europe, Middle East and Africa
ESL	English Skills Training
EU	European Union
FDI	Foreign Direct Investment
GDP	Gross Domestic Product
GGDC	Groningen Growth and Development Centre
GVCs	Global Value Chains
HEIs	Higher Education Institutions
HR	Human Resources
ICT	Information and Communication Technology
ICT-ITES	Information and Communication Technology-Information Technology Enabled Services.
ILO	International Labour Organization
IMF	International Monetary Fund
IoT	Internet of Things
ISIC Rev.3	Standard Industrial Classification of All Economic Activities
IT	Information Technology
ITE	Information and Technology Education
ITES	IT-enabled Services
ITO	Information Technology Outsourcing
KIBS	Knowledge-Intensive Business Services
KPO	Knowledge Process Outsourcing
KZN	KwaZulu Natal region

LPO	Legal Process Outsourcing
LQ	Location Quotient
LVCs	Local Value Chains
M&A	Mergers and Acquisitions
MNC	Multinational Company
NACE	Nomenclature statistique des activités économiques dans la Communauté européenne
NAFTA	North American Free Trade Agreement
NASSCOM	National Association of Software and Services Companies
NMC	New Middle Class
NUTS	Nomenclature of territorial units for statistics
NYSE	New York Stock Exchange
OECD	Organization for Economic Cooperation and Development
OLI	Ownership, Location and Internalisation
ORN	Offshoring Research Network
PWT 8.0	Penn World Table 8.0
R&D	Research and Development
RVCs	Regional Value Chains
SADC	Southern African Development Community
SLA	Service Level Agreement
SME	Small or Medium Enterprise
SSA	Sub-Saharan Africa
TiSA	Trade in Services Agreement
TFP	Total Factor Productivity
TP-BPO	Third-Party Business Process Outsourcing
UIS	UNESCO Institute for Statistics
UKTI	UK Trade and Investment
UNCTAD	United Nations Conference on Trade and Development
UNESCAP	United Nations Economic and Social Commission for Asia and the Pacific
WIOD	World Input-Output Database
WTO GATS	World Trade Organization the General Agreement on Trade in Services

Contributors

Niels Beerepoot is Assistant Professor in Economic Geography at University of Amsterdam. He holds an MSc. in International Economics and Economic Geography from Utrecht University and a PhD in Human Geography from the University of Amsterdam. His research interests include the local outcomes of economic globalisation in Southeast Asia, service offshoring and the rising middle class in emerging economies.

José Antonio Camacho is Lecturer in Economics at University of Granada and director of the Institute of Regional Development. His research interests include input-output analysis, service activities and regional development.

Mike Crone is a Senior Lecturer in International Business at Sheffield Business School, Sheffield Hallam University, UK. He previously worked at the University of Strathclyde, Glasgow, and Queen's University, Belfast, and before that at Northern Ireland Economic Research Centre. Mike has a PhD in Economic Geography from University of Sheffield, UK. His research interests lie at the interface of international business and economic geography. He has previously published in journals such as Regional Studies, Environment and Planning C: Government and Policy and European Planning Studies and has recently authored several book chapters.

Gisela Di Meglio is an Assistant Professor in the Department of Economic Analysis II at Complutense University of Madrid, Spain. She is also a Council member of the European Association for Research on Services (RESER). Previously, she has been a research fellow at the University of Alcalá, Spain; where she has obtained her PhD. Her research interests are: structural change, economic growth and development, productivity, competitiveness and efficiency of services.

Hilal Erkuş-Öztürk is an Associate Professor at Akdeniz University (Faculty of Economics and Political Sciences, Antalya, Turkey). She is specialized in urban and regional economic development, political economic geography, tourism development, tourism clusters, networks, tourism value chains, urban tourism governance, economic homogenization and diversification in tourism firms.

Robert Huggins is Chair of Economic Geography and Director of Research and Innovation at Cardiff University's School of Planning and Geography, as well as Director its Centre for Economic Geography. Professor Huggins's key areas of interest include regional economic development, in particular the study of competitiveness, culture, knowledge flows, entrepreneurship, innovation, and inter-organisational networks. He has published more than seventy articles in peer-reviewed journals, as well as numerous books.

Andrew Johnston is Principal Lecturer in International Business and Economics at Sheffield Business School, Sheffield Hallam University. He is currently leader of the International Business subject group as well as Director of the International Business and Economics Research Group (IBERG). His research interests include innovation, entrepreneurship, and regional development, particularly the role of universities in innovation and regional development and he has published widely in these fields.

Jonathan Jones is a Lecturer in Economics at Newcastle University Business School. His research interests are in empirical microeconomics applied specifically to foreign direct investment (FDI). His work has been published in a number of economics and regional science journals and is the co-author of a book on foreign direct investment and the regional economy. He has also been involved as an affiliate of the ESRC/UK government funded Spatial Economics Research Centre. His work has attracted media interest at both the national and regional level, including newspapers, BBC web-based coverage, and television and radio appearances.

Charlotte Keijser is a PhD fellow at MGSoG/UNU-MERIT. She holds a master's degree in International Development Studies from the University of Amsterdam (UvA) and a bachelor's degree in Human Geography. Charlotte has carried out research in India, Ghana, Togo, Senegal and South Africa. She joined the PhD programme in 2013. Her research interests include value chains of services and innovation systems in sub-Saharan Africa.

Jana Kleibert is a Post-doc researcher at the Leibniz Institute for Research on Society and Space (IRS) in Erkner, Germany, and is honorary research fellow at the School of Environment, Education and Development at the University of Manchester. Her research interests are global production networks, services offshoring and economic development in the Global South, in particular the Philippines. She holds a PhD in Economic Geography and a Master's degree in Political Science (International Relations) from the University of Amsterdam. She currently teaches at the Free University of Berlin.

Robert C. Kloosterman is a Professor of Economic Geography and Planning at the University of Amsterdam. He is Honorary Professor at the Bartlett School of Planning, University College, London and held the Franqui Chair in Entrepreneurship at the Faculty of Business Studies, Hasselt University

(Belgium) in 2012. He is interested in what kind of economic activities take place in contemporary urban environments and how they are related to other aspects such as the institutional framework, the historical legacy in terms of economic orientation and urban morphology.

Rosa Koetsenruijter is a Human Geography undergraduate and honours student at the University of Amsterdam. She is specialised in the field of economic geography. Other interests include political geography and environmental geography. As an intern at the Department of Human Geography, Urban Planning and International Development Studies, she worked in close collaboration with Robert C. Kloosterman and Niels Beerepoot.

Bart Lambregts lectures urban planning at the Faculty of Architecture of Kasetsart University in Bangkok. He holds an MSc. in Urban Planning from Delft University of Technology and a PhD in Human Geography from the University of Amsterdam. His current research concentrates on the role of services industries in processes of globalisation and local/regional development, sustainable urban mobility issues, and the interdependencies between urban development practices and climate resilience building.

Mary Leian Marasigan is Researcher at the School of Labour and Industrial Relations, University of the Philippines. She has published a number of researches on labour and workers' rights in the Philippines and in Asia, most recent of which on temporary employment and non-standard work in Indonesia, the Philippines and Japan. She is a PhD candidate at the University of Amsterdam where is she studying employability and offshore services in the Philippines.

H. Richard Nakamura is an Assistant Professor at the School of Business, Economics and Law, University of Gothenburg, Sweden. He was awarded a PhD from Stockholm School of Economics. His main research area is foreign direct investments and cross-border mergers and acquisitions (M&A), where he focuses on investment processes and managerial issues concerning multinational enterprises' international operations and asset ownership.

Panos Papadongonas holds a bachelor's degree in Human Geography and Urban and Regional Planning from University of Amsterdam and a master's degree in International Development Studies from the same university. His research interests include cultural and economic globalisation, the rising new middle class in emerging economies, community dynamics and collective identification processes in multicultural settings. He currently works as researcher focusing on value-based market segmentation.

Mercedes Rodríguez is Lecturer in Economics at the University of Granada where she received her PhD in 2003. She is also a member of the Institute of Regional Development. Her research interests focus on innovation activity in services, and in particular in knowledge-intensive business services (KIBS), and on the role of these activities in the growth process of emerging economies.

Patrik Ström is Associate Professor in Economic Geography, Centre for International Business Studies, Department of Business Administration, University of Gothenburg. He is also the Head of the Department of Business Administration. He holds a PhD in Business Administration from Roskilde University and an Econ. Dr. in Economic Geography from the University of Gothenburg. His research focuses on the internationalisation and competitiveness of the knowledge-intensive service industry in Europe and Asia. Ström is the president of the European Association for Research on Services (RESER) and he was also member of the EU Commission High Level Group on Business Services, where he chaired the working group on Internationalisation of Services.

Pieter Terhorst is a Lecturer at the University of Amsterdam (Department of Geography, Planning and International Development Studies, Amsterdam, The Netherlands). He is specialized in economic geography, variety of capitalism, tourism cities, clusters, rescaling of state spaces.

Robert Wentrup is Trade Commissioner and Country Manager at Business Sweden – The Swedish Trade and Invest Council in France. He is PhD candidate in economic geography at the Centre for International Business Studies in the Department of Business Administration, University of Gothenburg in Sweden. His research draws from economic geography, international business, and development studies, and is about how the Internet is altering the economic geography, and more specifically about the internationalisation pattern of online service providers. He has a background as a consultant in Accenture's technology practice.

Colin Wren is a Professor in Economics at Newcastle University. His research is in regional economics, focusing on the interface between government and industry, with implications for economic development. Recent work examines industrial location, and in particular foreign direct investment. He has carried out many studies evaluating the impact of public interventions and has served on international and national bodies. He was a project leader on a recent ESRC Spatial Economics Research Centre and past editor of Regional Studies. He has published numerous articles in mainstream Economics and field journals, and authored four books on economic development and related issues.

Acknowledgements

This collection of writings on *Globalisation and Services-driven Economic Growth* finds its origin in the mini-conference that was held at the University of Amsterdam on 19 and 20 June, 2014. This event took place under the banners of the Commission on the Dynamics of Economic Space of the International Geographical Union, and the Regional Science Association International. The Commission on the Dynamics of Economic Space has a long and rewarding tradition in organising small, thematically focused conferences. We would like to thank Neil Reid and Christine Tamásy, as former and current chair of the Commission, for their support for this book project and we are pleased that another book can be added to the commission's book series – the 11th, if our calculations are right!

We are thankful for the financial support that we received for the mini-conference from the International Geographical Union. Furthermore, we acknowledge the support of the Netherlands Organisation for Scientific Research-Science for Global Development (NWO-WOTRO) whose research grant (grant number W 01.65.329.00) enabled free time for compiling this book.

We thank the contributing authors for their enthusiasm and support to this book project and their timely response to all our queries. In particular we thank Pieter Terhorst for his advice and detailed feedback on various chapters. In addition we thank Rosa Koetsenruijter for her assistance in preparing the manuscript. Finally, we thank Katy Crossan and Faye Leerink as our commissioning editors at Ashgate and Routledge for their encouragement and pleasant cooperation.

Niels Beerepoot, Bart Lambregts and Jana Kleibert

1 Globalisation and services-driven economic growth

An introduction

Bart Lambregts, Niels Beerepoot and Jana Kleibert

The ongoing shift to services and the globalisation of services production

Over the past several decades the weight and the role of services in the global economy have steadily gained importance. Originally, services were primarily seen as a residual category of economic output that could not be attributed to either agriculture or manufacturing. The more or less continuous growth of the sector throughout the twentieth and early stages of the twenty-first century however, has led to a gradual reappraisal (Daniels, 2004; Elfring, 1989). While interest in the role of services in economic transformation goes back to at least halfway the previous century (see for example Clark, 1940; Kuznets, 1957; Stigler, 1956), it has intensified notably from the 1970s onwards, spurred by developments such as the rapid growth of producer services and the heightened recognition of producer services' role in innovation and competitiveness (Hertog, 2000; OECD, 2000; Wood, 2002a). More recently still, the digitisation and advances in information and communication technologies (ICT) that in combination with services trade liberalisation have allowed for the unbundling and drastic reorganisation of services production have fuelled further debate (Baldwin, 2011; Blinder, 2006; Lambregts *et al.*, 2015). According to the latest estimates, services account for no less than 62.5 per cent of global GDP and about 43 per cent of global employment (CIA Factbook, 2015). Moreover, a wide and rapidly growing variety of services has become tradeable in the past decades, enabling countries across the world, including in the Global South, to bank on services exports as an economic growth strategy (UNCTAD, 2004). Services now feature firmly alongside commodities and manufactured goods in global and regional trade agreements and their potential role in development has started to gain the full attention from academics as well as international development agencies such as the World Bank, UNCTAD, UNESCAP, IMF, ILO, OECD and ADB. Services, in sum, have evolved from a dependent and largely neglected economic category to the world's largest and arguably most dynamic generator of value added and jobs (ILO, 2015), a major driving force of globalisation (Blinder, 2006, Bryson, 2007; Dossani and Kenney, 2007), and – possibly – the next engine for growth in countries across the Global South (Ghani and O'Connell, 2014; Kleibert, 2015;

Lambregts *et al.*, 2015). This naturally makes the world of services an increasingly relevant field of interdisciplinary scholarly enquiry, with a number of issues vying for attention. This book concentrates on the question how the global reorganisation of services production alters the relation between and generates different sets of challenges and opportunities in the Global North and the Global South. It examines and provides empirical accounts of how changes in the geography of service delivery generate new interdependencies between services producing and services consuming regions across the globe; how services help to mitigate the impact of and contribute to recovery from economic crises in the Global North; and how the unbundling and relocation of services production fosters economic development and service sector-driven modernisation processes in new locations in the Global South. The remainder of this introductory chapter further sets the scene, identifies some of the major issues at stake, defines the book's objective and outlines the rest of the book.

The ongoing shift towards services

The world of services, as noted, is rapidly changing. For considerable time services have been overlooked as central to the global economy. Services were regarded not to be capable of driving economic growth since labour productivity in services tended to grow slower than in manufacturing (Baumol, 1985) and production and consumption took place in close proximity to each other, so limiting the potential for trade in services. However, with more types of services becoming tradeable and notably advanced business services having developed the capacity to create substantial added value and spread productivity gains across an economy, this view has gradually changed. Globally, the growth of employment in private sector services has outperformed growth in manufacturing for many years and it is expected to continue to do so for years to come (ILO, 2015). Today, in most countries of the Global North the share in GDP of producer and consumer services combined is between 70 and 80 per cent (World Bank, 2015a). Beyond the Global North, the importance of services has risen dramatically as well, especially in the most globally connected metropolises of the Global South and in the Asia-Pacific region. Here, knowledge-intensive producer services (Daniels and Harrington, 2007; Taylor, 2004; Zhang, 2015) as well as more generic but highly tradeable low-end business services (Kleibert, 2015; Lambregts *et al.*, 2015) have found fertile grounds shaped by a combination of booming markets for high- and medium-end producer services and favourable production conditions for low-end business services.

The rise of services over the past decades can be linked to two key driving forces. The first and perhaps most obvious concerns the rise in disposable income enjoyed by many – first in the Global North and now increasingly also in the Global South – and generated by ongoing labour productivity gains in agriculture and manufacturing. Initially, in the first couple of post-WWII decades, a substantial part of that income was spent on household appliances, actually reducing the demand for various personal services (Gershuny, 1978; Gershuny and Miles,

1983). However, in due time and in line with classic economic development thinking (Kuznets, 1973) other categories of services started to benefit, and some of them even benefitted greatly. Between 1960 and 1987 the share of services in total employment in France, Germany, Japan, the Netherlands, Sweden, the United Kingdom and the United States increased from 46 to 65 per cent, with notably social services (healthcare and education in particular) and the then still smallish producer services sector growing fast (Elfring, 1988). This trend broadly continued into the new millennium (OECD, 2000; 2015a), with producer services and health and education being joined by the leisure and hospitality sector as a major contributor to services employment growth (BLS, 2015). The latter, together with producer services (but now excluding financial services) so far have also shown the strongest signs of recovery after the 2008 financial crisis (BLS, 2015; OECD, 2015a), notwithstanding the fact that the same crisis – certainly in the Global North – has put pressure on real wage growth (OECD, 2015a) and thus households' financial ability to purchase services.

A second important driver of services growth relates to the production system in general and the environment in which it is embedded having become more complex and subject to rapid change. This has stimulated demand for transactional, managerial, analytical, consulting and design activities in production processes, and boosted the growth – again first in the Global North and now increasingly also in the Global South – of an increasingly versatile and knowledge-intensive producer services industry (Castells, 1996; Daniels, 1993; Sassen, 1991; Wood, 2002b). An additional source of growth more recently has presented itself in the form of IT-enabled outsourcing and offshoring of services from the Global North to low-wage economies in the Global South. This trend (discussed in more detail below) keeps the production cost of routinized but tradeable services in check and allows for continued delivery of low value-added services that at Western wage levels would no longer be economically viable. For instance, labour-intensive and low value-added services activities such as the digitisation of paper archives, certain types of transcription work and certain types of customer relations management probably would not be produced (anymore) or had become fully automated if offshoring to low-wage countries was not an option. The rise, even more recently, of online platforms for the trade of micro-services (i.e. small to very small electronically transferable services tasks) is likely to make even more services tasks economically viable and thus to further expand the market for services (Lehdonvirta and Ernkvist, 2011). Platforms such as *Upwork* and *Freelancer* enjoy increasing popularity and if this trend continues its impact on the world of services production may well become tangible (Beerepoot and Lambregts, 2015; World Bank, 2015b).

In the process, services have come to represent an enormously variegated array of activities. The diversity of services, Castells argued, is 'in fact so extreme that it forbids considering services as a single, homogeneous sector of economic activity' (1989: 129). Services, however, can be classified into different categories. For starters there is the distinction between 'traditional' and 'modern' services (Bryson, 2007), where traditional services are defined as requiring face-to-face interaction tying the time and location of production to that of consumption. A classic example of a traditional service is a haircut or a restaurant meal – examples of a modern service would be the translation of a document or the development of an app, which can be conducted over distance. A second common distinction is between consumer services and producer (or business) services. The latter are intermediate inputs into further production activities and usually (but not exclusively) transacted among and between firms and organisations. Consumer (or personal) services in contrast are services sold to individuals or organisations for final consumption (OECD, 2000). A third distinction is between market and non-market services, where the former are services produced for (profitable) sale on the market and the latter provided to the public at no or reduced costs. Yet another way of categorising services is by the amount of knowledge that goes into the production of the service (UNCTAD, 2004). This produces the distinction between high-end, knowledge-intensive (business) services (KI(B)S), and low-end or routinised services at the other end of the scale. The former for instance include R&D, financial, legal and engineering services while the latter contain for example services related to administration, customer care and data processing services. Another useful distinction still is between tradeable and non-tradeable services, with the former covering all services that can be traded across space and the latter referring to those services for which production and consumption necessarily coincide in time and space. The different typologies show different degrees of overlap and frequently are used indiscriminately. None of them comprehensively grasps the diversity and complex nature of services alone.

Further complicating matters is that the distinction between manufacturing and services as two distinct spheres is increasingly blurred (Daniels, 2004; Zysman *et al.*, 2013). Some activities categorised as services revolve around processes that are essentially industrial: think, for instance, restaurants, where the kitchen sees natural resources being transformed into man-made, higher value commodities. Many services activities meanwhile remain statistically hidden in non-services sectors, as about every production chain in agriculture and manufacturing also includes services activities. It is estimated that services, as intermediate inputs, contribute to more than 30 per cent of value added in manufactured goods (OECD-WTO-UNCTAD, 2013: 18). In manufacturing exports, services value added accounts for almost a third in developed countries and 26 per cent in developing economies (Lanz and Maurer, 2015). Such services, however, only become visible in statistics and hence 'formally' part of the services domain when they are outsourced to companies specialised in the delivery of such services. At least a part of the remarkable *recorded* growth in services is actually precisely because of that: business processes such as payroll management, customer care, recruitment, market analysis, advertising and so on being outsourced to third party services providers and hence becoming statistically recorded as services. Yet, the other way round it is manufacturing that the services industry depends upon for much of its productivity growth. Many routine services tasks are rapidly being replaced by increasingly capable machines or intelligent algorithms (Brynjolfsson and McAfee, 2014). The ongoing mechanisation and automation of services production makes that ever more of such processes are in fact taking on the characteristics of industrial production.

The globalisation of services production

Outsourcing and, in close connection, offshoring of services activities underlie much of the current dynamics visible in the services domain. Following drastic shifts in the (spatial) organisation of the production of goods (Fröbel *et al.*, 1980), increasingly fierce competition now forces firms to also look critically at how the production of services is organised. Advances in ICTs have enabled firms to unbundle service business processes, and the lowering of trade barriers in combination with increased global availability of sufficiently skilled labour allow for the relocation of evermore of these processes to there where they can be performed in the most cost-efficient way (Eichengreen and Gupta, 2013; Gereffi and Fernandez-Stark, 2010a). The codification, standardisation, and digitisation of information have enabled the production of services in one place, and consumption in another, allowing for transnational trade in services (UNCTAD, 2004). Tradeable services now travel the world faster and at lower transportation costs than physical goods have ever done.

Since its take-off two decades ago, the offshoring of services production has taken a great leap. In result, a new geography of services production is taking shape: a geography that is defined by new interregional and international divisions of labour and held together by increasingly complex global services production networks. The geographical changes are profound. When, before the advent of modern ICTs, production and consumption necessarily still took place in close proximity, the spatial distribution of services production closely followed that of population and economic activity in general. Over time, with the industry spawning evermore variegated arrays of consumer and producer oriented services and with modern ICTs and trade liberalisation allowing an increasing number of especially producer oriented services to be traded across space, a much more varied location pattern and a new pattern of specialisation in services production has developed. Today, high-end producer services (as well as high-end consumer services) are concentrated predominantly in the primary cities. Simultaneously, a dispersion of lower-end services has occurred (Dicken, 2015; Sassen, 1991). Whereas such lower-end (back-office) services initially relocated to cheaper and more accessible places in the periphery of large cities and to lower-cost peripheral regions in the Global North (Garreau, 1992; Richardson *et al.*, 2000), they are currently being moved to places offering the right mix of location factors in the Global South. Much of the work lands in large cities such as Bangalore, Mumbai, Manila and Shanghai. However, in major receiving countries such as India and the Philippines, already a further dispersal of offshored services activities into secondary and even tertiary cities can be observed (Beerepoot and Vogelzang, 2015; Kleibert, 2014; Chapter 8 this volume). Meanwhile, the shift towards a tertiary sector-dominated economy has led to increasing polarisation in developed economies. In US cities, for example, but also in cities such as Singapore and Hong Kong, high-paid cognitive-cultural workers coexist with a low-paid underclass of service employees (Castells, 1989). In the Global South the emergence of sizeable offshore services economies in cities such as Bangalore, Mumbai, Manila and Delhi fuels the expansion of the local middle class, or, as some prefer: the rise of a *new* middle class (Fuller and Narasimhan, 2007; Murphy, 2011; Upadhya, 2008). Similar to cities in the Global North, polarisation here is observed to be increasing as well (Krishna and Nederveen-Pieterse, 2008).

Some scholars believe that the impact of these trends on the global economy will be larger than the earlier relocation of manufacturing activities (e.g. Dossani and Kenney, 2007). Services offshoring is believed to present new opportunities for developing countries and possibly enable some to leapfrog industrialisation and become service-based economies (Dossani and Kenney, 2007; Gereffi and Fernandez-Stark, 2010b; Ghani and O'Connell, 2014; Kloosterman *et al.*, 2015). Probably the most well-known case is India's transformation into the global back-office of the world, delivering information technology and other services to the Global North. Ghani (2010: 3) argues that 'the South Asian experience suggests that a services revolution – rapid growth and poverty reduction led by services – is possible. This can fundamentally change the pattern of development for many developing countries'. The local outcomes of services offshoring in developing countries, however, are diverse (see e.g. Hernández *et al.*, 2014; Lambregts *et al.*, 2015) and the potential for developing countries to be uplifted by pursuing 'export-oriented tertiarisation' of their economies, remains hotly debated. The Indian economists Chandrasekhar and Gosh (2006), for example, argue that IT services are an 'exaggerated development opportunity' since they are unable to provide enough employment opportunities for unskilled workers. Others warn that the industry is extremely footloose and that for any benefitting region or country the risk of employment and derivative development effects being short-lived is high (Beerepoot *et al.*, 2015; UNCTAD, 2004). Rodrik (2014) in turn warns that the increasing share services make up in developing countries' economies should not uncritically be read as a sign of services-led development but rather a result of manufacturing experiencing 'premature' decline (Rodrik, 2015). He warns that many developing countries do not yet offer the right conditions, notably as regards the quality of labour supply, for rapid and sustained growth of skill-intensive, high-productivity and high-wage services sectors (which indeed could function as growth escalators) and sees clear limitations to productivity growth in non-tradeable services depending on domestic demand as growth of the latter largely depends on productivity gains in the rest of the economy. Rodrik is sceptical therefore 'that a services-led model can deliver rapid growth and good jobs in the way that manufacturing once did' (2014: 3).

For the Global North one key question is whether offshoring of relatively low-value-added tasks creates opportunities for labour to engage in higher-value-added activities and (thus) strengthens the (global) competitiveness of local firms, or whether the trend primarily adds to local unemployment problems and real income decline (Bardhan *et al.,* 2013, Breznitz and Zysman 2013,

Milberg and Winkler 2013, Rubalcaba, 2007). Optimists maintain that the cost benefits enjoyed by firms that engage in offshoring will foster the creation of better, more productive jobs 'at home' and so lift the productivity level of the entire economy. Others, in contrast, think that the process will not nearly yield enough high-productive jobs to replace all the jobs offshored and see most of the people affected move to lower value added (non-tradeable) services jobs instead. With producer and business services (representing higher productivity jobs) and leisure and hospitality services (representing lower productivity jobs) in the US having added comparable numbers of jobs between 2000 and 2014 (2.4 million versus 2.8 million; BLS, 2015), it is probably too early to name the winner. A perhaps bigger question still is the old but persistent doubt about whether services also in the long term can function as the (global) economy's key engine of growth (Petit, 1986; Rodrik, 2015). While services in the Global North have contributed greatly to employment and value added, productivity growth in services has remained slow and in many countries slower than in manufacturing and other industries (OECD, 2005). If the economies of the Global North wish to enjoy similar economic growth rates as they did prior to the 2008 financial crisis, productivity growth in services has to increase (ibid.). How this is to be achieved while maintaining employment at the same time, is a challenge extraordinaire.

Aim and focus of the book

This book engages with the broader debate about the changing role of services in the global economy. According to Ghani and O'Connell (2014: 2–3) current developments in services call for a rethink of at least three long-cherished tenets of economic development. First, increased tradability of specific services undermines the long-held believe that services are driven by domestic demand and as such cannot by themselves drive growth. With foreign demand now capable of driving service growth, export-oriented services can now be local drivers of growth. Second, services were long considered to have lower productivity than manufactures. However, with services production having become more IT and knowledge-intensive and with more services gaining importance as inputs in manufacturing processes (i.e. the rise of producer services) the productivity gap between manufacturing and services, while not closed yet, has significantly decreased, rendering a reconsideration of this image timely. Third, it was long thought that 'good' jobs were created only in the manufacturing sector. Today's services sector, however, is certainly not devoid of stable, well-paid and fulfilling jobs, perhaps even more so than the manufacturing sector itself. If we add to this that global trade in service now grows faster than trade in goods and that service-led growth is compatible with greener, inclusive, and more gender friendly growth (ibid.), but also that informed concerns remain about the long-term capacity of services to generate productivity and economic growth, the promise held and challenges produced by the shift towards services becomes abundantly clear.

It is against this backdrop that the book aims at understanding how the global reorganisation of services production alters the relation between and generates different sets of challenges and opportunities in the Global North and the Global South. To this end it examines: a) how trade in services creates new interdependencies between services producing and services consuming regions across the globe; b) how services help to mitigate the impact of and contribute to recovery from economic crises in the Global North; and c) how the unbundling and relocation of services production fosters economic development and service sector-driven modernisation processes in new locations in the Global South. These constitute the three main thematic foci of the book and are elaborated below.

Services and globalisation

Services have risen in importance both for developed and developing countries. New patterns of global production, often functionally disintegrated and spatially separated, connect distant regions in the Global North and Global South through a 'trade in tasks' (Grossman and Rossi-Hansberg, 2008). The effects on places should therefore not be read separately, but as profoundly interrelated implications of the globalisation of services that affect regions, cities and people around the globe. Services perform an important function as intermediaries in the global economy, but can also constitute global production networks by themselves (Coe and Yeung, 2015). Business services and ICT enabled services are important lubricants in the global economy. In some sectors, the distinction between manufacturing and services becomes blurred, as services take on progressively more important functions as inputs in manufacturing activities (Zysman *et al.*, 2013). The new digital economy makes increasing use of globally-tradeable business services which are sourced from evermore distant locations. Mapping global trade flows in advanced business services is therefore important to grasp the geographies of global trade in services (see Rodríquez and Camacho, this volume). While evermore locations become integrated and connected to the digital economy, the access and use of ICTs remains fundamentally uneven (Graham *et al.*, 2012; 2014). The new and changing geographies of trade in services require therefore further analysis (see Wentrup *et al.*, this volume).

The changing role of services in the Global North

In post-industrial societies in the Global North the services sector is important for employment generation and as a generator of value added (with a key role for producer services in the latter). Especially in times of economic crisis, services, many of which are characterised by relatively low entry barriers in terms of capital investments and regulatory requirements, are an important avenue for resilience and employment creation in disadvantaged regions. For instance, new business start-ups in Europe since the outbreak of the financial crisis in 2008 – a crisis co-produced ironically by an important exponent of the services sector itself – have been predominantly in the broad services sector (OECD, 2015b), with substantial numbers of people trying to make a living as self-employed, freelance personal or business services providers (Hatfield, 2015). Advanced regions in the Global North meanwhile have to ensure their viability though increasing productivity and competitiveness in high-end services. Creativity and cognitive-cultural activities play an increasingly important role in this (Scott, 2012, see also Kloosterman and Koetsenruijter in this volume). The challenge ahead in the Global North is to accommodate the impacts that corporate restructuring and increasing tradability of services production have on employment and economic recovery. Breznitz and Zysman (2013) in this respect pose the question of how developed economies can stay competitive in 'the third globalisation', the most recent phase of economic restructuring. When assessing the opportunities for regional economic development resulting from services, the location choice patterns of investments are important indicators for where employment opportunities are created. Jones and Wren (this volume) show that in the context of the United Kingdom, foreign direct investments (FDI) in services have overtaken manufacturing FDI. They further show that these exhibit different location trends from manufacturing. The knowledge-based economy is predominantly rooted in urban regions. Simultaneously, economically deprived regions in the Global North can be imagined to be severely affected by increasing geographic flexibility of services employment. Yet, they can also benefit from new economic opportunities. One such case is Northern Ireland, which has attracted sizeable offshore service investments over the past years, as Crone (this volume) shows. Meanwhile, the importance of both traditional and modern consumer services in maintaining employment levels and urban attractiveness should not be ignored. Many of these can also be classified into higher-end and lower-end services, with judgement often based on subjective evaluations of quality. Terhorst and Erkuş-Öztürk (this volume) explore how uncertainty pertaining to quality assessments in such markets is reduced, using the restaurants scene in Amsterdam as an example. All things considered, the implications for growth, competitiveness and resilience are central to the discussion of the impact of services in the Global North.

Service-led modernisation in the Global South

The rising importance of services employment and its effects on local economies requires further attention. An important question is whether the Global South can become the stage of viable service sector-driven growth. Services have played a vital role in recent economic growth in Asia (ADB, 2012), but also countries in Latin America and Africa attract foreign direct investment in services. Services offshoring is usually accompanied by optimism on the potential positive effects it can have on countries in the Global South, including skills development and trickle-down effects through increased spending and consumption. In close connection, questions arise about how services provide employment opportunities for marginalised people, what role specific places play in attracting these services, and how service firms can enter into new markets.

The outcomes of the shifting patterns of service delivery and consumption in the Global South are beginning to become mapped (Bardhan *et al.*, 2013; Hernández *et al.*, 2014; Kleibert, 2015; Lambregts *et al.*, 2015). What becomes clear is that the developmental outcomes are fundamentally empirical questions requiring further analysis. Overly optimistic claims of services being capable of creating economic development miracles would seem as misplaced as defeatist, race-to-the-bottom-like arguments that focus on labour exploitation and other downsides alone. Instead, a combination of qualitative and quantitative analysis is required to generate a more comprehensive and balanced understanding of the effects of services on growth and economic development in the developing world. We for instance observe that South African information technology-enabled services (ITES) firms not only create new North-South linkages, but also regional, intra-African delivery chains and networks. While such intra-South networks are often neglected (see e.g. Mans (2014), for a critique), they change the geography of service delivery in various ways (Keijser, this volume). But also other transformations resulting from the offshoring and outsourcing of ITES services require scrutiny. Are new skills developed as a result of the offshoring of lower-end services for labour arbitrage (see Beerepoot and Hendriks, 2013; Marasigan and Lambregts, this volume)? Are new services sectors stimulated through the relatively high incomes earned by workers in the offshore service sector, such as the local retail sector, allowing for broader participation and employment opportunities (see Papadongonas and Beerepoot, this volume)?

Overview of the book

The chapters in this book concentrate on the three thematic foci outlined above: services and globalisation; the changing role of services in the Global North; and service-led modernisation in the Global South. Different chapters deal with different types of services, with the three principal types of services investigated being advanced business services, information technology-enabled services (ITES) and consumer services. Some chapters take an explicitly geographic approach and examine changes in the spatial distribution of services production or consumption in different parts of the world. Other chapters investigate the inner workings of the service sector by examining quality uncertainty in service markets and giving empirical accounts of how quality upgrading in service delivery takes place and how services offshoring affects human capital development in the Global South. The chapters combined use a wide variety of qualitative and quantitative methods and present a wealth of original empirical research material. An overview is provided in Table 1.1.

In Chapter 2, Mercedes Rodríguez and José Antonio Camacho empirically assess the evolution, between 1995 and 2011, of the global sourcing of business services inputs drawing on data provided by the World Input-Output Database. Now that firms can effectively access business services suppliers across the world, there exists in fact a global market for business services. The authors show how a new global pattern of trade flows in business services inputs is emerging,

Table 1.1 Chapter summary

Chapter	*Theoretical discussion*	*Type of services*	*Empirical data*	*Geographical focus*
Part I: Services and globalisation				
2	Globalisation of business services	Advanced business services	World Input-Output Database information on business services	Trading blocs (Eurozone, other EU, NAFTA, China, East Asia, BRIIAT)
3	Kaldorian framework analysis of services as growth engine	Tertiary sector with special focus on business services	GGDC 10-Sector Database – Africa Sector Database-Penn World Table	Latin America (9 countries), Africa (11 countries), Asia (9 countries)
4	Global connectivity and online service provision	Information technology enabled online consumer services such as Twitter, Facebook, Wikipedia	Quantitative data on internet penetration (World Bank), largest online service providers, Facebook usage	Global patterns but zooming in on Africa for studying the geography of Facebook use
Part II: The changing role of services in the Global North				
5	The role of the service sector in economic resilience	Advanced business services, consumption based services	Quantitative data on employment and regional GDP at NUTS 2 level	268 NUTS 2 regions in European Union
6	Trade flows and locational patterns in service FDI	Financial services, software and computer services, R&D, business services (NACE classification)	NUTS 2 level data on inward FDI in Great Britain	Great Britain
7	Quality and information uncertainty in consumer service markets	Consumer services (restaurants)	Qualitative data by means of interviews with 37 restaurateurs and 50 tourists	Amsterdam (Netherlands)
8	Service offshoring and regional development in recipient locations	Advanced business services	Qualitative and quantitative secondary data	Northern Ireland

Table 1.1 continued

Chapter	*Theoretical discussion*	*Type of services*	*Empirical data*	*Geographical focus*
9	Role and potential of cultural industries for development	Four cultural industries: music production, film, fashion, architectural design	Secondary data on performance cultural industries	BRIC countries, East Asian Tigers (Hong Kong, Singapore, Taiwan, South Korea), United States, United Kingdom
Part III: Service-led modernisation in the Global South				
10	Growing significance of regional value chains in the ITES sector	Information Technology enabled Services (producer services)	Qualitative data by means of interviews with stakeholders in the South African ITES sector	South Africa
11	Skills utilisation and development in offshore services production	Information Technology enabled Services (producer services)	Quantitative (survey) data and qualitative interview data	Manila (Philippines)
12	Indirect effects of service sector driven growth	Information Technology enabled Services, consumer services (retail sector)	Qualitative data by means of interviews with retailers and workers in the ITES sector	Mumbai (India)

largely driven by the rapid relocation of business services to India and China from the 1990s on and the subsequent expansion into the rest of Asia, Eastern Europe and Latin and Central America in the 2000s.

Gisela di Meglio (Chapter 3) explores the link between the service sector, economic growth and development to identify whether some service activities can function as an engine of economic growth in developing countries. Kaldor's growth model is used to analyse changes in the composition of production and employment, and its interaction with patterns of sectoral productivity growth. The chapter finds diverse paths of sectoral transformation between the developed and the developing world in which a rapid emergence of services is taking place along with a declining participation of industry.

In Chapter 4, Robert Wentrup, Richard Nakamura and Patrik Ström examine the spatial inequalities of internet services by analysing the usage of online services (demand side) and the provision of these services by focusing on the geographical origin and internationalisation of the world's largest online service providers at the firm level. They observe that most of the demand and supply of online services comes from the high-end urban districts in the Global North showing that there is still an important digital divide in demand and supply.

Andrew Johnston and Robert Huggins (Chapter 5) concentrate on the contribution of knowledge-based services to economic development with a particular focus on the years following the 2008 economic crisis. They examine changes in service sector-employment across European NUTS 2 regions and find that in the period between 2008 and 2012 employment grew faster in business services than in consumer services. The results provide support for the argument that business services play a major role in economic resilience and that it is therefore important for policymakers not to ignore the potential of knowledge-intensive service sectors in their regional development programmes.

Also focusing on business services, Jonathan Jones and Colin Wren (Chapter 6) analyse the spatial distribution of foreign direct investment in services at the NUTS 2 level for Great Britain over the period 1996–2005. Foreign direct investment in services has been the main form of new inward investment in Great Britain since the late 1990s when it overtook incoming manufacturing FDI. Financial services, software and computer services, R&D and business services are responsible for 70 per cent of FDI projects during this period. Remarkably, service FDI tends to be overrepresented in the main economic centre (London and South-East England) and more peripheral areas (Wales, North-East England) of Great Britain signalling a new spatial division of labour in which the existing economic centre attracts higher-end services and lower-end business services (e.g. call centres) settle in peripheral areas.

Shifting towards consumer services, Pieter Terhorst and Hilal Erkuş-Öztürk (Chapter 7) argue how quality and information uncertainty lead to distortions in the functioning of service markets. Taking the case study of the restaurant sector in Amsterdam, they wonder how certain segments in this market can survive under conditions of quality uncertainty. In practice, quality standards are largely set locally as well as nationally, and, to a lesser extent, globally. This also matters in other service markets where the quality of the (intangible) product can only be evaluated after its consumption or where objective criteria for quality are missing. Quality uncertainty in service markets requires mediators who set quality standards and help to reduce information asymmetry.

Mike Crone (Chapter 8) shows how since the mid-1990s Northern Ireland, a region in the 'old periphery' of North-West Europe, has become an attractive destination for offshore service-related FDI. Peripheral regions of developed economies (like Northern Ireland) have been largely overlooked in recent research on service offshoring which has tended to focus on relocation to the Global South (exemplified in this book by Chapters 10–12). Similar to locations in the Global South, concerns are raised about whether the location of branch-plants of international service firms provides a new form of dependent development. The case study also shows that work continues to disperse to smaller second tier cities in order to benefit from lower operating cost.

Cultural industries (e.g. music production, film, fashion and architectural design) constitute a growing subsector within services. Robert Kloosterman and Rosa Koetsenruijter (Chapter 9) refute the commonly held assumption – based on the premise that higher levels of income and higher levels of purchasing power generate a higher demand for cultural products – of a relationship between the stage of economic development and the size and significance of a country's cultural industries. They show how both the BRICS and some East Asian countries have become important players in the changing global division of labour in the production of cultural products. The issue for the future is whether they can contest the existing Western hegemony in these activities and shape a new global geography in cultural industries.

Examining advanced business services in the Global South, Charlotte Keijser (Chapter 10) concentrates on the position of South Africa as service provider in the ITES market. While South Africa is a relatively small player in the global market for offshore services, it plays a major role in the Sub-Saharan regional market for service delivery. The chapter highlights the growing significance of non-Western markets for service provision. In the South African case, the domestic and regional ITES market allowed some firms to build capacity and gain experience to compete in the global market place.

In Chapter 11, Leian Marasigan and Bart Lambregts engage with the question whether and how the arrival of offshored service work provides opportunities for skill upgrading in the receiving country's labour force. They examine how the ITES sector appeals to the skills of the workers and whether and how workers' skill portfolios are enhanced through working in the industry. They find that Metro Manila's offshore services industry generally underutilises the skills of its mostly university educated workers but that working in the industry does enable workers to add specific skills to their skill portfolios.

In Chapter 12, Panos Papadongonas and Niels Beerepoot investigate how the rise of the ITES sector in Mumbai impacts on the local consumer service market. The growth of the ITES sector added to the formation of a 'new middle class' with considerable spending power, which reshapes the local market for luxury consumer goods. Not only does the expansion of the ITES sector stimulate growth of the retail sector, it also leads to new perceptions among its workers on luxury consumption and social status. The rise of advanced business services is accompanied by quality upgrading in the local retail sector, hereby demonstrating the indirect effects of service-led growth.

In the book's final chapter the editors synthesise the cases, identify the key message emerging from them, and signal directions for future research.

References

Asian Development Bank. (2012). *Asian Development Outlook 2012 Update. Services and Asia's Future Growth.* Mandaluyong City: Asian Development Bank.

Baldwin, R. (2011). *Trade and Industrialisation after Globalisation's 2nd Unbundling: How Building and Joining a Supply Chain are Different and Why it Matters* (No.

w17716). Washington: National Bureau of Economic Research. Available at: www.nber.org/papers/w17716 [accessed on 21 October 2015].

Bardhan, A., Kroll, C. A. and Jaffee, D. M. (eds) (2013). *The Oxford Handbook of Offshoring and Global Employment*. Oxford University Press.

Baumol, W. J. (1985). Productivity growth and the service sector. In: R. P. Inman (1988). *Managing the Service Sector: Prospects and Problems*. Cambridge: Cambridge University Press.

Beerepoot, N. and Hendriks, M. (2013). Employabilty of offshore service workers in the Philippines: opportunities for upward labour mobility or dead-end jobs? *Work, Employment and Society* 27(5), pp. 823–41.

Beerepoot, N. and Lambregts, B. (2015). Competition in online job marketplaces: towards a global labour market for service outsourcing? *Global Networks* 15(2), pp. 236–55.

Beerepoot, N., Lambregts, B. and Kloosterman, R. C. (2015). Conclusions: offshore services and the road to development. In: B. Lambregts, N. Beerepoot and R. C. Kloosterman (eds) *The Local Impact of Globalization in South and Southeast Asia: Offshore Business Processes in Services Industries*. London and New York: Routledge, pp. 208–14.

Beerepoot, N. and Vogelzang, E. (2015). Service outsourcing to smaller cities in the Philippines. The formation of an emerging local middle class. In: B. Lambregts, N. Beerepoot and R. C. Kloosterman (eds) *The Local Impact of Globalization in South and Southeast Asia: Offshore Business Processes in Services Industries*. London and New York: Routledge, pp. 195–207.

Blinder, A.S. (2006). Offshoring: the next industrial revolution? *Foreign Affairs* 85(2), pp. 113–28.

Breznitz, D., and Zysman, J. (2013). *The Third Globalization. Can Wealthy Countries Stay Rich in the Twenty-First Century?* Oxford and New York: Oxford University Press.

Brynjolfsson, E and McAfee, A. (2014). *The Second Machine Age: Work, Progress and Prosperity in a Time of Brilliant Technologies*. New York: W.W. Norton & Company.

Bryson, J. (2007). The second global shift: The offshoring or global sourcing of corporate services and the rise of distanciated emotional labour. *Geografiska Annaler: Series B, Human Geograph* 89(S1), pp. 31–43.

Bureau of Labor Statistics (BLS) (2015). Employees on nonfarm payrolls by industry sector and selected industry detail, various years. [Online] Available at: www.bls.gov/webapps/legacy/cesbtab1.htm [accessed on 8 November 2015].

Castells, M. (1989). *The Informational City: Information Technology, Economic Restructuring, and the Urban-Regional Process*. Oxford, UK and Cambridge, MA: Basil Blackwell.

Castells, M. (1996). *The Rise of the Network Society*. Oxford: Blackwell.

Chandrasekhar, C. and Ghosh, J. (2006). IT-driven offshoring: The exaggerated 'development opportunity'. *Human Systems Management* 25(2), 91–101.

CIA (2015). The World Factbook. [Online] Available at: www.cia.gov/library/publications/the-world-factbook/geos/xx.html [accessed on 22 October 2015].

Clark, C. (1940). *The Conditions of Economic Progress*. London: Macmillan.

Coe, N. M. and Yeung, H. W.-C. (2015). *Global Production Networks. Theorizing economic development in an interconnected world*. Oxford: Oxford University Press.

Daniels, P. W. (1993). *Services Industries in the World Economy*. Oxford: Blackwell.

Daniels, P. W. (2004). Reflection on the 'old' economy, 'new' economy, and services. *Growth and Change* 35(2), 115–137.

Daniels, P. W. and Harrington, J. W. (eds) (2007). *Services and Economic Development in the Asia-Pacific*. Ashgate Publishing: Aldershot.

Dicken, P. (2015). *Global Shift: Mapping the Changing Contours of the World Economy* (7th edition). London: SAGE Publications.
Dossani, R. and Kenney, M. (2007). The next wave of globalization: Relocating service provision to India. *World Development* 35(5), pp. 772–91.
Eichengreen, B. and Gupta, P. (2013). The two waves of service-sector growth, *Oxford Economic Papers* 65(1), pp. 96–123.
Elfring, T. (1988). Service employment in advanced economies, a comparative analysis of its implications for economic growth. Groningen: Dissertation Universiteit Groningen.
Elfring, T. (1989). New evidence on the expansion of service employment in advanced economies. *Review of Income and Wealth* 35(4), pp. 409–40.
Fröbel, F., Heinrichs, J. and Kreye, O. (1980). *The New International Division of Labour.* Cambridge: Cambridge University Press.
Fuller, C.J. and Narasimhan, H. (2007). Information technology professionals and the new rich middle class in Chennai (Madras). *Modern Asian Studies* 41(1), pp. 121–50.
Garreau, J. (1992). *Edge City: Life on the New Frontier* (1st Anchor Books ed.). New York: Anchor Books.
Gereffi, G. and Fernandez-Stark, K. (2010a). The Offshore Services Global Value Chain. North Carolina: CORFO-Report, Center on Globalization, Governance and Competitiveness. Durham, NC: Duke University.
Gereffi, G. and Fernandez-Stark, K. (2010b). The offshore services value chain: Developing countries and the crisis. In O. Cattaneo, G. Gereffi, and C. Staritz, *Global Value Chains in a Postcrisis World.* Washington, DC: World Bank, pp. 335–372.
Gershuny, J. (1978). *After Industrial Society: The Emerging Self-Service Economy.* London: Macmillan.
Gershuny, J. and Miles, I. (1983). *The New Service Economy: The Transformatoin of Employment in Industrial Societies.* New York: Praeger.
Ghani, E. (2010). *The Service Revolution in South Asia.* Oxford University Press.
Ghani, E. and O'Connell, S.D. (2014). Can Service Be a Growth Escalator in Low Income Countries? World Bank Policy ResearchWorking Paper 6971,Washington, DC:World Bank.
Graham, M., Hale, S. and Stephens, M. (2012). Digital divide: The geography of internet access. *Environment and Planning A* 44(5), pp. 1009–1010.
Graham, M., Hogan, B., Straumann, R.K. and Medhat, A. (2014). Uneven geographies of user-generated information: patterns of increasing informational poverty. *Annals of the Association of American Geographers* 104(4), pp. 746–64.
Grossman, G. M. and Esteban Rossi-Hansberg, E. (2008). Trading tasks: A simple theory of offshoring. *American Economic Review* 98(5), pp. 1978–97.
Hatfield, I. (2015). Self-employment in Europe. London: Institute for Public Policy Research (IPPR).
Hernández, R.A., Mulder, N., Fernandez-Stark, K., Sauvé, P., López Giral, D. and Muñoz Navia, F. (eds) (2014). *Latin America's Emergence in Global Services. A new driver of structural change in the region?* Santiago, Chile: United Nations publication.
Hertog, P. den (2000). Knowledge-intensive business services as co-producers of innovation. *International Journal of Innovation Management* 4(04), pp. 491–528.
International Labour Organization (ILO) (2015). World Employment and Social Outlook – Trends 2015. Geneva: International Labour Office.
Kleibert, J.M. (2014). Strategic coupling in 'next wave cities': Local institutional actors and the offshore service sector in the Philippines, *Singapore Journal of Tropical Geography* 35(2), pp. 245–60.

Kleibert, J.M. (2015). Expanding global production networks: The emergence, evolution and the developmental impact of the offshore service sector in the Philippines. Amsterdam: Dissertation Universiteit van Amsterdam.

Kloosterman, R. C., Beerepoot, N. and Lambregts, B. (2015). Service-sector driven economic development from a historical perspective. In: Lambregts, B., Beerepoot, N. and Kloosterman, R. C. (eds) *The Local Impact of Globalization in South and Southeast Asia: Offshore Business Processes in Services Industries*. London and New York: Routledge, pp. 17–28.

Krishna, A. and Nederveen-Pieterse, J. (2008). Hierarchical integration: The dollar economy and the rupee economy. *Development and Change* 39(2), pp. 219–37.

Kuznets, S. (1957). Quantiative aspects of the economic growth of nations: II. Industrial distribution of national product and labor force. *Economic Development and Cultural Change* 5(4), pp. 1–111.

Kuznets, S. (1973). *Modern Economic Growth, Rate, Structure, and Spread.* New Haven/London: Yale University Press.

Lambregts, B., Beerepoot, N. and Kloosterman, R. C. (eds) (2015). *The Local Impact of Globalization in South and Southeast Asia: Offshore Business Processes in Services Industries.* London and New York: Routledge.

Lanz, R. and Maurer, A. (2015). Services and global value chains – some evidence on servicification of manufacturing and service networks. WTO Working Paper ERSD-2015-03, World Trade Organization, Economic Research and Statistics Division. Available at: www.wto.org/english/res_e/reser_e/ersd201503_e.pdf [accessed on 22 November 2015].

Lehdonvirta, V. and Ernkvist, D. (2011). *Knowledge Map of the Virtual Economy: Converting the Virtual Economy into Development Potential*, Washington: IBRD.

Mans, U. (2014). Understanding the varieties of green-driven growth: Cities and renewable energy in the Global South. Amsterdam: Dissertation Universiteit van Amsterdam.

Milberg, W. and Winkler, D. (2013). *Outsourcing Economics: Global Value Chains in Capitalist Development.* Cambridge: Cambridge University Press.

Murphy, J. (2011). Indian Call Centre Workers: Vanguard of a global middle class? *Work, Employment and Society* 25(3), pp. 417–33.

OECD (2000). OECD Employment Outlook. Paris: OECD.

OECD (2005). Growth in Services. Fostering Employment, Productivity and Innovation. Paris: OECD.

OECD (2015a). OECD Employment Outlook. Paris: OECD.

OECD (2015b). Entrepreneurship at a glance 2015. Paris: OECD.

OECD-WTO-UNCTAD. (2013). *Implications of Global Value Chains for Trade, Investment, Development and Jobs.* Report prepared for the G-20 Leaders Summit: www.oecd.org/trade/G20-Global-Value-Chains-2013.pdf.

Petit, P. (1986). *Slow Growth and the Services Economy.* London: Frances Pinter.

Richardson, R., Belt, V. and Marshall, N. (2000). Taking calls to Newcastle: the regional implications of the growth in call centres. *Regional Studies* 34(4), pp. 357–69.

Rodrik, D. (2014, October 13). *Are services the New Manufactures?* [Online] Available at Project Syndicate: www.project-syndicate.org/commentary/are-services-the-new-manufactures-by-dani-rodrik-2014-10 [accessed on 21 October 2015].

Rodrik, D. (2015). Premature Industrialisation. NBER Working Paper Series (No.20935) Washington: National Bureau of Economic Research. Available at: www.nber.org/papers/w20935 [accessed on 10 November 2015].

Rubalcaba, L. (2007). *The New Service Economy – Challenges and Policy Implications for Europe.* Cheltenham: Edward Elgar.

Sassen, S. (1991). *The Global City: New York, London, Tokyo*. Princeton, NJ: Princeton University Press.

Scott, A. J. (2012). *A World in Emergence. Cities and Regions in the 21st Century.* Cheltenham, UK and Northampton, MA, USA: Edward Elgar.

Stigler, G. J. (1956). Trends in Employment in the Service Industries. *National Bureau of Economic Analysis*, 59, General Series. National Bureau of Economic Research.

Taylor, P. J. (2004). *World City Network: A Global Urban Analysis*. London: Routledge.

UNCTAD. (2004). *World Investment Report 2004: The Shift Towards Services*. New York and Geneva: United Nations Conference on Trade and Development.

Upadhya, C. (2008). Rewriting the Code: Software Professionals and the Reconstitution of Indian Middle Class Identity. In: Jaffrelot, C. and Van der Veer, P. *Patterns of Middle Class Consumption in India and China*, pp. 55–87. New Delhi: Sage.

Wood, P. (2002a). Knowledge-intensive services and urban innovativeness, *Urban Studies* 39, pp. 993–1002.

Wood, P. (2002b). *Consultancy and Innovation: The Business Service Revolution in Europe*. London: Routledge.

World Bank (2015a). World Development Indicators. [Online] Available at: http://data.worldbank.org/indicator/NV.SRV.TETC.ZS [accessed on 22 February 2015].

World Bank (2015b). *The Global Opportunity in Online Outsourcing*. Washington, DC: World Bank.

Zhang, X. (2015). *From the 'Workshop of the World' to an Emerging Global City-Region: Restructuring of the Pearl River Delta in the Advanced Services Economy*. Amsterdam: Dissertation Universiteit van Amsterdam.

Zysman, J., Feldman, S., Kushida, K. E., Murray, J. and Nielsen, N. C. (2013). Services with everything: the ICT-enabled digital transformation of services. In: Breznitz, D. and Zysman, J. *The Third Globalization* pp. 99–129. Oxford: Oxford University Press.

Part I

Services and globalisation

2 Mapping world trade flows in business services

Mercedes Rodríguez and José Antonio Camacho

Introduction

Trade in business services has been growing rapidly over the last decades both in developed and developing countries (Goswami *et al.*, 2011). The development of new technologies and the subsequent falling in communication costs, in combination with trade liberalisation, have driven the potential for trade in a world with an ever more internationally fragmented production. This is reflected in the fact that most international trade is related to goods and services used as inputs in production processes. Thus, nowadays more than half of world goods imports are used as inputs and the percentage is superior in the case of services: more than 70 per cent of world services imports are used as inputs (Backer and Miroudot, 2013). In the particular case of business services the scope for growth of trade is vast as they are among the 'more tradable' industries. For instance, according to Jensen and Kletzer (2010) about 70 per cent of total employment in business services in the United States could be traded. However, it is also necessary to note that as recent analyses highlight (Van der Marel and Shepherd, 2013), trade in business services is comparatively more sensitive to policy restrictiveness than the rest of services. Special mention deserves restrictions to trade in professional services.

One of the main problems in analysing recent trends in trade in services is the lack of adequate statistics. Currently there are only two main surveys which contain firm-level data on outsourcing and offshoring of services: the Offshoring Research Network (ORN) Research Survey launched in 2004 at Duke University and the outsourcing surveys conducted by Statistics Denmark in 2007 and 2012. At the macro-level international databases do not normally contain disaggregated data for exported and imported services used as inputs. To our knowledge, the World Input-Output Database (WIOD) constitutes the first published input-output database that allows, not only to differentiate between domestic and imported business services inputs in each country or region, but also to identify the country or region of origin of those imported business services inputs. The WIOD was constructed within a project funded by the European Commission Research Directorate General as part of the 7th Framework Programme, Theme 8: Socio-Economic Sciences and Humanities. The project ran from May 1, 2009 to May 1,

2012. Its main purpose was to increase the understanding of increasing global integration through international trade and technological development. The database contains a set of harmonised supply and use tables and data on international trade in goods and services.

The objective of this chapter is to empirically assess the evolution of the global sourcing of business services inputs during the period 1995–2011 drawing on the information provided by the WIOD. The chapter is structured as follows. In the next section the existing literature on the global sourcing of business services is reviewed. The third section contains a brief description of the WIOD and the empirical analysis. The fourth section looks at bilateral trade flows and the concluding section summarises the main findings and discusses some possible future analyses.

Are business services inputs going global?

As Bhagwati (1984) predicted in his pioneering work on trade in services, the development of new technologies and internet usage has radically affected the tradability of most business services (Abramowsky and Griffith, 2006; Freund and Weinhold, 2002) to such an extent that national advantages in new technologies are an important source of competitiveness in business services (Guerrieri and Meliciani, 2005). Blinder (2006: 114) notes that 'because packets of digitalized information play the role that boxes used to play, many more services are now tradable'. New technologies do not only alter the geography of low-skilled services provision but also the geography of high-skilled ones. Some authors go further and sustain that new technologies are having a levelling effect, giving place to the emergence of a 'flat world' (Friedman, 2005). However, looking more carefully at this issue using economic measures it becomes clear that rather than flat the world seems to be 'spiky' and 'in terms of both sheer economic horse-power and cutting edge innovations surprisingly few regions truly matter in today´s global economy.' (Florida, 2005: 48). This can be explained by the fact that much of the knowledge that is valuable for firms is tacit and requires personal contact and trust to adequately transfer. Thus, when knowledge-intensive activities are traded cultural and linguistic barriers emerge (Malecki, 2010) and, as a result, socio-institutional aspects become key elements to explain whether knowledge can be shared over long distances (Maskell and Malmberg, 1999; Gertler, 1995, 2003).

In spite of these potential difficulties, a recent strand of literature points out that 'knowledge service clusters' are emerging in developing countries, defined as 'geographic concentrations of lower-cost technical and analytical skills serving rising global demand for commoditized knowledge services.' (Manning, 2013: 379). This implies a radical change in the long dominant vision of economic geography that established a clear distinction between the type of activities that cluster in developed and emerging countries, namely, high-tech activities in the case of developed countries and low-cost ones in emerging countries (Mudambi, 2008). The most cited example is India, and more concretely the

Bangalore software industry (Arora and Gambardella, 2005; Chaminade and Vang, 2008; Parthasarathy and Aoyama, 2006), but there are many other cities in China, Eastern Europe or Latin America that are becoming key providers of knowledge services (Manning, *et al.* 2010).

What are the main advantages of global sourcing for firms? In their review of the international business literature Cheung *et al.* (2008) describe four major explanatory factors for the positive impact that the international sourcing of services has on productivity: firms can specialise and focus on their core activities, it may help to business restructuring, it may free up resources that can be invested in productivity-enhancing capital or technology and, in the case of knowledge-intensive services, firms can learn from their providers. They conclude that the international sourcing of service inputs may have a greater impact on productivity than material inputs: as over the past two decades the marginal benefit of material offshoring has declined considerably while the international sourcing of services is a newer phenomenon and has much more room to grow.

The potential benefits pointed out above encourage firms to fragment production and use business services produced abroad. However, to adequately analyse how internationalisation patterns change over time it is necessary to couple the traditional ownership, location and internalisation (OLI) advantages described by Dunning and Norman (1983) with the strategy (Faulconbridge *et al.*, 2008; Taylor *et al.*, 2013; Taylor *et al.*, 2014) because the internationalisation of business services firms 'is integral to their strategy' (Faulconbridge *et al.*, 2008: 211). In a recent study on global value chains conducted by Backer and Miroudot (2013) they retake the distinction established by Gereffi and Fernandez-Stark (2010) between horizontal and vertical services to demonstrate that not only the use, but also the production of business services, is becoming more and more global. Horizontal business services are defined as those services activities needed by any type of firm (software, research and development, accounting, legal services, and so on) while vertical business services are those business services that are part of a specific value chain. Given the available statistical information, they focus on two groups of horizontal business services: computer and related activities and other business services. In the case of other business services the participation of countries like India, the United Kingdom and Belgium in the global value chains is particularly high but there are differences across these countries in terms of position. Thus, while the United Kingdom or Belgium tend to be specialised in services rather upstream in India the specialisation is more downstream. This fact is in line with the insights revealed by the literature on advanced producer services and world city networks: 'while a firm may have tens or hundreds of offices, not all of these offices share an equal level of strategic importance' (Taylor *et al.*, 2014: 271).

As we mentioned before, a complementary debate focuses on whether global value chains are really 'global' or rather 'regional'. Although some authors sustain that regional blocs still play a key role (Johnson and Noguera, 2012; Baldwin and López-González, 2013), the study conducted by Los *et al.* (2015) using data from the WIOD finds a stronger pace of growth for international fragmentation than for regional fragmentation, or, in other words, that belonging to a trade bloc still matters but less than some decades ago. We pointed out in the introduction that trade in business services is particularly sensitive to trade restrictions. Jensen (2013) cites as the most significant examples of barriers to trade in business services the requirements on the form of establishments, foreign partnerships restrictions, ownership and investment restrictions, nationality requirements, residence and local presence requirements, licencing and accreditation of foreign professionals, limitations on the scope of activities, multi-disciplinary practices restrictions and fee and advertising restrictions. The current negotiation of the Trade in Services Agreement (TiSA) tries to address two main types of barriers to trade in services: limitations to national treatment, that is, treating foreign suppliers differently from local ones, and limitations to market access, that is, the extent to which foreign suppliers can operate. Nonetheless, 'whether interest will translate into serious engagement is not clear.' (Wolfe, 2015: 22).

As for the positive impact of trade in business services inputs at the macro-level, empirical studies conducted both in developed and developing countries support the hypothesis put forward by Markusen more than 25 years ago: as business services inputs are differentiated and complementary to domestic production, permitting free trade in these services has a positive effect on productivity and innovation (Markusen, 1989). In the case of developed countries the papers by Amiti and Wei (2005, 2009) for the United States show that the use imported service inputs has a positive effect on productivity in the manufacturing sector using industry-level data. Similar studies for other countries such as Germany (Winkler, 2010) or Belgium (Michel and Rycx, 2014) reach the same result: the use of business services produced abroad leads to productivity gains in manufacturing. The evidence using firm-level data is scarcer but we can highlight some works conducted for Ireland (Görg and Hanley, 2005; Görg *et al.*, 2008) or Japan (Hijzen *et al.*, 2010).

In emerging economies the greater availability of business services also plays a very positive role, not only to increase productivity but also to foster innovation. For instance, different studies for India (Arnold *et al.*, 2012), China (Zhang *et al.*, 2010), Russia (Jensen *et al.*, 2007; Knobel, 2012) or the Czech Republic (Arnold *et al.*, 2011) point out that services reforms, and in particular trade liberalisation, give access to better and more diverse business services and lead to improvements in productivity and, in most cases, in innovation.

In brief, in this chapter we focus on trade in business services inputs because, as Jensen and Kletzer (2005) have noted, they are not only higher skilled and higher paying than manufacturing but also higher skilled and higher paying than non-tradable services. Second, in contrast to some pessimistic visions, liberalisation policies benefiting foreign providers and the subsequent increase in the number of foreign firms can positively affect real wages of domestic skilled workers (Markusen *et al.*, 2005).

Data and empirical analysis

Data

The service sector is very heterogeneous. While geographers commonly employ the term 'producer services' most economists use 'business services'. Anyway, as Daniels (2001) highlights, the existence of very different classifications makes it particularly difficult to undertake reliable and comparative analyses of the development of these activities. Miles *et al.* (1995) distinguish two main types of business services: those featuring physical functions (like transport) and those providing information and knowledge functions (like computer services, research and development and consultancy). The WIOD lists the following four industries of the International Standard Industrial Classification of All Economic Activities (ISIC Rev.3) as business services: 71 Renting of machinery and equipment without operator and of personal and household goods; 72 Computer and related activities; 73 Research and development and 74 Other business activities. This division comprises four main groups of services: legal, accounting, bookkeeping and auditing activities; tax consultancy; market research and public opinion polling; business and management consultancy; architectural, engineering, and other technical activities; advertising and business activities not included elsewhere (n.e.c.). In other words, most of business services included within this classification are knowledge-intensive business services.

The WIOD covers 40 countries: the EU27, Canada, United States, Brazil, Mexico, China, India, Japan, South Korea, Australia, Taiwan, Turkey, Indonesia and Russia. In order to facilitate the analyses, the 40 countries are grouped into six major regions: the Eurozone, the Other EU, the NAFTA, China, East Asia and the BRIIAT. We have to mention that in the case of the EU countries they are classified as Eurozone members or non-euro EU members as of 1 January 2007. The countries included within each region are shown in Table 2.1.

The main advantage of the WIOD, in addition to being publicly available, is that it provides time series of global, regional and national input-output tables. In particular the last edition covers the years from 1995 to 2011. All national tables are harmonised which permits the construction of regional and world tables by combining the national tables with bilateral trade international data. This is particularly interesting for services, as the level of detail of trade in services statistics is usually not as high as in manufacturing.

Our empirical analysis in conducted in two phases. First, we examine the evolution of the use and trade flows in business services inputs by region. Next, we analyse bilateral trade flows.

Global trade flows in business services inputs

In Figure 2.1 we provide shares of business services inputs in total output and the percentage of those business services inputs imported. We plot the shares in 1995 and the annual average growth rates over the period 1995–2011.

Table 2.1 Regional aggregation in world input-output tables

Eurozone	*Other EU*	*NAFTA*	*China*	*East Asia*	*BRIIAT*
Austria	Bulgaria	Canada	China	Japan	Australia
Belgium	Czech Republic	Mexico		Korea	Brazil
Cyprus	Denmark	USA		Taiwan	India
Estonia	Hungary				Indonesia
Finland	Latvia				Russia
France	Lithuania				Turkey
Germany	Poland				
Greece	Romania				
Ireland	Sweden				
Italy	UK				
Luxembourg					
Malta					
Netherlands					
Portugal					
Slovakia					
Slovenia					
Spain					

Source: Timmer (2012: 42).

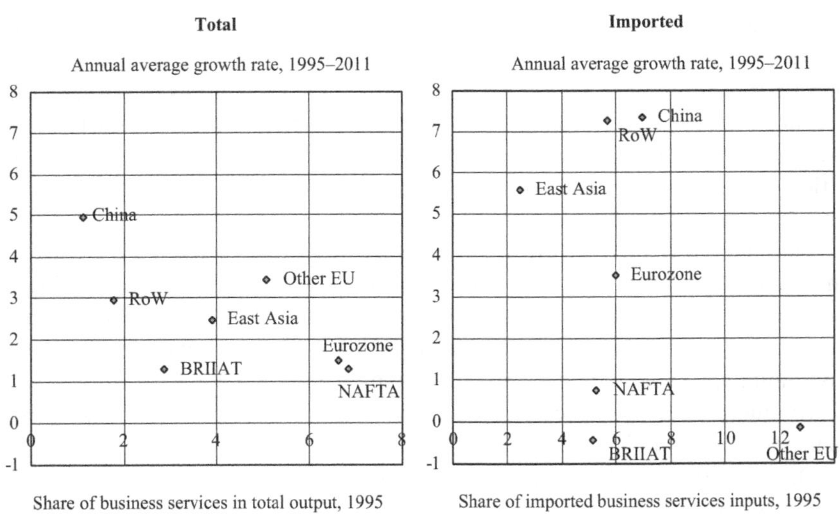

Figure 2.1 Evolution of the use of business services inputs, 1995–2011

Source: WIOD.

Despite marked variations in the share of business services inputs in total output (in 2011 the share exceeded 8 per cent in the EU countries and the NAFTA while in China was 2.35 per cent) all regions registered an intensive growth in the use of business services inputs during the period analysed. In some cases, like in China and, to a lesser extent, in East Asia, an important part of this growth relied on imports. The rise in imported business services inputs was also noticeable in the Eurozone. It is remarkable that, in spite of the similar intensity in the use of business services as inputs, the percentage business services inputs imported in the EU was more than double the percentage of those imported in the NAFTA in 2011 (of total business services employed as inputs 10.25 per cent in the Eurozone and 12 per cent in Other EU countries were imported compared to only 5.90 per cent in the NAFTA). Among the emerging economies East Asia consolidates as a more intensive user of business services compared to the BRIIAT or China. In terms of foreign sourcing China appears as the most dependent region on imports whereas the percentage of imported business services inputs were almost identical in East Asia and the BRIIAT (4.75 per cent and 4.62 per cent of total business services inputs, respectively, in 2011).

The computation of normalised trade balances can provide a first glance at the regions' competitiveness in trade in business services inputs. In Figure 2.2 we provide the normalised trade balances for the years 1995 and 2011 in order to find out which regions gained or lost competitiveness.

As can be seen, in 2011 the greatest deficit was reported by East Asia while the BRIIAT registered the highest surplus. This latter region along with China experienced the most radical change over the period analysed, moving from deficits of around 22 per cent in 1995 to a surplus of 21.22 per cent and 2.46 per cent, respectively, in 2011. All the 'high-intensive users', namely, the EU and the NAFTA, improved their trade balances although there were no major changes in terms of balance: the NAFTA continues to report a deficit substantially higher than the Eurozone while the Other EU countries show a noticeable surplus.

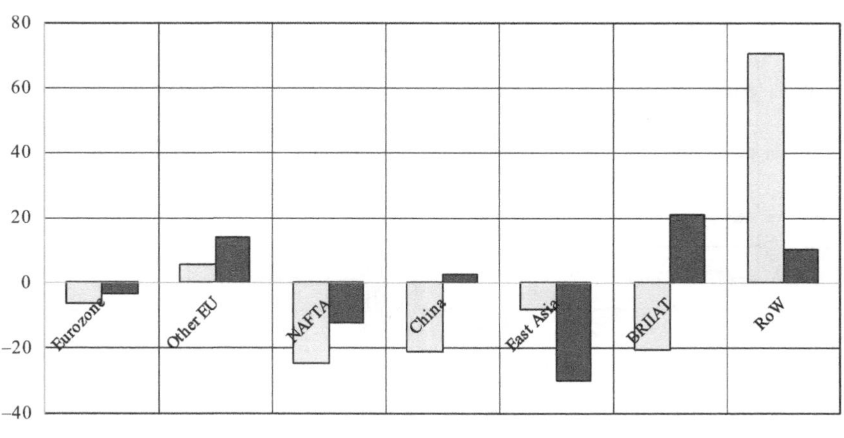

Figure 2.2 Normalised trade balances in business services inputs, 1995–2011

Source: WIOD.

In brief, while only the Other EU countries where net exporters in the beginning of the period, in 2011 the BRIIAT and China also registered positive balances in their trade in business services inputs. As was pointed out above, the NAFTA and the Eurozone were net importers in 2011 but unlike East Asia, where the deficit grew at a fast pace, they reduced their deficits over the period. Obviously, India has much to say in the positive evolution experienced by the BRIIAT: according to the World Bank (2004) India experienced the fastest growth in services exports in the world over the 1990s. In the Chinese case its participation in global value chains has differed from that of other Asian countries as it has been able to tap into advanced technology that was not available in the country and shift from simple assembly to controlling processes and product design. As a result, China is assuming an increasing role in knowledge-intensive segments of global value chains (OECD, 2013).

Finally, before entering into the analysis of bilateral trade flows, we examine the global position of each region in world trade in business services inputs in Figure 2.3. As can be observed, the Eurozone continues to be the world's leading business services inputs exporter, with a share close to 30 per cent, followed by the NAFTA. The Other EU countries were closely behind, with quite a similar share in global exports (17.8 per cent compared to 18.4 per cent in the NAFTA). Together these three regions which were classified above as 'high-intensive users' accounted for more than 65 per cent of world exports in business services inputs in 2011.

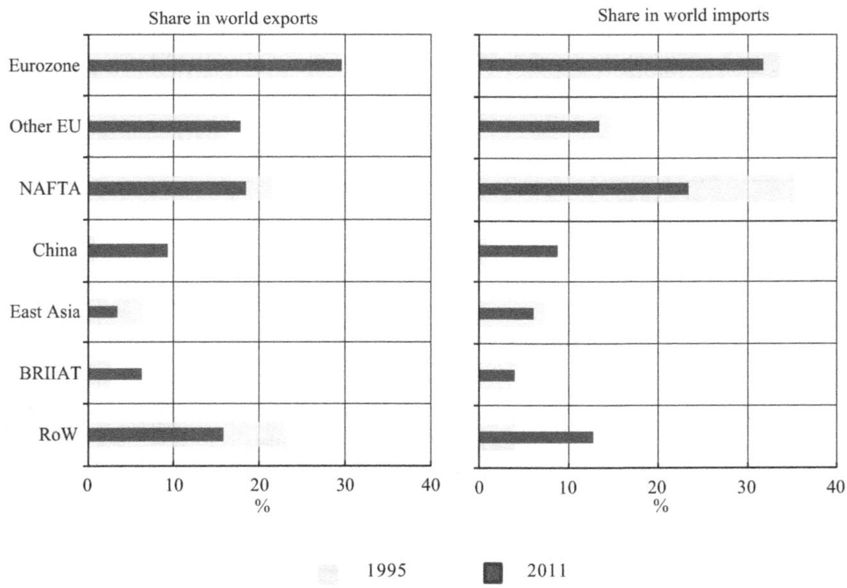

Figure 2.3 Share in world trade in business services inputs, 1995–2011

Source: WIOD.

It is necessary to highlight, however, the differing trends followed by the EU and the NAFTA: while the global position of the EU in terms of exports and imports remained more or less unaltered over the period, in the NAFTA there was substantial drop, particularly strong in the case of imports. Thus, although the NAFTA still was the second largest exporter in 2011, its share in global imports significantly decreased (more than 12 percentage points) over the period 1995–2011. The fact that domestic value added content is significantly higher in the case of business services exports in combination with the existence of few global leading suppliers (the US, Germany, Japan or the United Kingdom) can help to explain this trend. Thus, the domestic content of business services exports in 2009 was substantially superior in the NAFTA (reaching 96 per cent in the US and 93 per cent in Canada and Mexico) than in the EU countries (where the shares of domestic value added in services exports were below 90 per cent excepting only France, Germany and the United Kingdom) (OECD, 2013).

Again, the most striking change was shown by China, which became the fourth largest world exporter and importer of business services inputs. Concerning exports it deserves special mention that East Asia and the BRIIAT followed the opposite trends: while the market share of East Asia halved over the period in the case of the BRIIAT it doubled.

Bilateral trade flows in business services inputs

Now that the main changes in global trade flows have been described we focus on bilateral trade relations. Table 2.2 provides an overview of bilateral trade flows among the regions in 1995 and 2011: by columns the share of the different countries in total exports and imports of business services inputs are shown.

Concerning the leading trader, the Eurozone, there was a substantial decrease in the percentage of exports directed to the NAFTA and Other EU countries. Altogether, they accounted for more than 78 per cent of total Eurozone exports in 1995 while this share was slightly above 45 per cent in 2011, largely thanks to the growing importance of emerging economies, in particular China, which has become a major client. Thus, China accounted for more than 14 per cent of total Eurozone exports in 2011. As for imports, the NAFTA and the Other EU countries continued to account for more than half of the Eurozone imports. It is necessary to note, however, that while the NAFTA lost share in the Eurozone the Other EU countries gained more than ten percentage points. A similar exchange in the shares in imports is observed between East Asia and the BRIIAT: while the share of East Asia in total Eurozone imports halved over the period, it doubled in the case of the BRIIAT.

As for the group of Other EU countries, the most striking change in terms of exports was the severe relative decline in those directed to the NAFTA: the share of the NAFTA in Other EU countries exports decreased from 60.95 per cent in 1995 to 27.55 per cent in 2011. As a result, the Eurozone was the most important market for Other EU countries business services inputs exports in 2011. In contrast, there was a severe decrease in imports from the Eurozone of 20 percentage points and, as

Table 2.2 Bilateral trade flows in business services inputs, 1995–2011

	Percentage of exports													
				1995							*2011*			
Country	*Eurozone*	*Other EU*	*NAFTA*	*China*	*East Asia*	*BRIIAT*	*RoW*	*Eurozone*	*Other EU*	*NAFTA*	*China*	*East Asia*	*BRIIAT*	*RoW*
Eurozone		32.94	57.64	17.32	46.04	53.71	49.68		45.65	44.83	12.28	39.07	46.42	63.71
Other EU	24.96		19.54	20.43	12.56	9.96	8.31	13.83		22.13	5.33	10.47	12.48	23.31
NAFTA	53.71	60.95		55.83	28.02	16.04	30.27	31.57	27.55		65.13	34.12	25.29	2.84
China	1.10	0.32	0.33		3.29	0.11	1.38	14.69	5.96	12.06		1.54	7.25	4.02
East Asia	7.75	3.42	10.67	5.37		10.15	8.25	4.39	3.93	11.83	12.21		5.02	3.36
BRIIAT	8.56	1.31	1.56	1.05	7.45		2.11	7.10	1.47	3.57	4.87	2.29		2.77
Row	3.91	1.06	10.26	0.00	2.65	10.03		28.42	15.44	5.58	0.19	12.51	3.53	

	Percentage of imports													
				1995							*2011*			
Country	*Eurozone*	*Other EU*	*NAFTA*	*China*	*East Asia*	*BRIIAT*	*RoW*	*Eurozone*	*Other EU*	*NAFTA*	*China*	*East Asia*	*BRIIAT*	*RoW*
Eurozone		50.35	44.93	33.31	31.22	62.69	29.17		30.42	39.92	49.60	21.13	52.75	65.45
Other EU	16.03		28.20	5.42	7.61	5.32	4.39	25.71		21.06	12.17	11.45	6.59	21.49
NAFTA	36.77	28.56		7.33	31.15	8.29	55.49	26.00	30.30		25.34	35.47	16.51	7.99
China	0.33	0.88	1.00		0.46	0.16	0.00	3.56	3.64	25.63		18.30	11.26	0.14
East Asia	8.56	5.35	4.95	21.09		11.51	4.18	4.10	2.59	4.86	0.58		1.92	3.25
BRIIAT	4.28	1.82	1.21	0.30	3.70		6.78	8.96	5.69	6.62	5.07	5.01		1.68
Row	34.04	13.04	19.71	32.54	25.86	12.03		31.67	27.36	1.91	7.24	8.64	10.98	

Source: WIOD.

happened in this latter region, China gained importance as trading partner. In terms of imports the BRIIAT region increased its share while East Asia lost relevance as provider.

Concerning the NAFTA, both exports and imports directed to and coming from the Eurozone experienced a decline. In contrast, China became a major trading partner, especially in terms of imports: one quarter of business services inputs came from China in 2011. In the case of the Other EU countries, although there was a substantial decline in the share of imports coming from this region, it still has a large market share in the NAFTA exports.

In China the NAFTA had an impressive export market share of 55.83 per cent in 1995 which grew ten percentage points over the period. The increasing role of China as trading partner for the NAFTA also sticks out when we look at imports, although we have to note that the main provider for China continues to be the Eurozone: almost half of the total business services inputs imported in China come from this region. The rise in imports coming from Other EU countries was also noticeable.

It is necessary to highlight the change experienced by East Asia: from being a key provider of business services inputs in 1995, its share was almost negligible in 2011. On the contrary, it became a very relevant destination for Chinese exports with a share very similar to the Eurozone.

As for East Asia, there was a drop in their trade flows with the EU (only the share of imports coming from Other EU countries increased over the period) while the flows with the NAFTA grew at a modest pace. In terms of imports, as mentioned above, the most relevant change was the increase in the share of business services inputs coming from China. Finally, regarding the BRIIAT we noted before the fast pace of growth experienced by exports. Nowadays, the Eurozone is the major client for the BRIIAT, followed by the NAFTA. The trends experienced by the imports directed to these two regions were, however, quite different. Thus, while the share of imports from the Eurozone substantially dropped the share of imports from the NAFTA doubled. Again, it is necessary to highlight the importance acquired by China and the diminution in the share of East Asia.

The changes in the origin and destination of trade flows in business services inputs translate into modifications in the sign of the trade balances among regions. In Table 2.3 we report the bilateral normalised trade among regions in 1995 and 2011. In the Eurozone, as a result of rising imports and decreasing exports to Other EU countries, the surplus registered in 1995 transformed in a substantial deficit in 2011. The same change in sign occurred in the trade balance with the BRIIAT, although in this case mainly due to the rapid increase in imports coming from this region.

As for the NAFTA, exports decreased at a faster pace than imports, so the surplus halved over the period. In 2011 China was not only the second major client for the Eurozone but it was also the region with which the Euro-zoned recorded its highest surplus.

Except for the NAFTA, the Other EU countries had trade deficits with all regions in 1995. However, over the period analysed they improved their trade

Table 2.3 Bilateral normalised trade balances in business services inputs, 1995–2011

	Percentage of exports													
	1995							*2011*						
Country	*Eurozone*	*Other EU*	*NAFTA*	*China*	*East Asia*	*BRIIAT*	*RoW*	*Eurozone*	*Other EU*	*NAFTA*	*China*	*East Asia*	*BRIIAT*	*RoW*
Eurozone		–0.16	–0.12	–0.49	0.11	–0.28	0.82		0.33	–0.06	–0.59	0.00	0.15	0.09
Other EU	0.16		–0.41	0.42	0.17	0.11	0.83	–0.33		–0.09	–0.37	–0.34	0.49	0.14
NAFTA	0.12	0.41		0.66	–0.13	0.12	0.52	0.06	0.09		0.46	–0.31	0.40	–0.39
China	0.49	–0.42	–0.66		0.72	–0.39	1.00	0.59	0.37	–0.46		–0.91	0.00	0.95
East Asia	–0.11	–0.17	0.13	–0.72		–0.26	0.84	0.00	0.34	0.31	0.91		0.60	0.12
BRIIAT	0.28	–0.11	–0.12	0.39	0.26		0.29	–0.15	–0.49	–0.40	0.00	–0.60		0.34
Row	–0.82	–0.83	–0.52	–1.00	–0.84	–0.29		–0.09	–0.14	0.39	–0.95	–0.12	–0.34	

Source: WIOD.

balances with all regions except the BRIIAT, where there was a substantial rise in the deficit, and the NAFTA, where there was a high decrease in the surplus.

Similarly to the Other EU countries, the NAFTA also registered deficits with all regions except East Asia in 1995. As a consequence of the good behaviour of exports, the trade balances with all regions improved over the period analysed with one sole exception, the BRIIAT. However, in spite of this good behaviour, and in difference with the Other EU countries, the NAFTA still presents trade deficits with all regions except East Asia. In China we can see how the negative balance with the Eurozone deteriorated over the period. The deterioration was even worse in the case of Other EU countries, where the surplus turned into deficit. The opposite evolution was found in the case of the trade balance with East Asia, where the high deficit registered in the beginning of the period turned into a great surplus in 2011.

The deterioration of the leading position of East Asia as a global provider is reflected in the fact that the surplus registered in its trade balance with all regions (except the NAFTA) in 1995 turned into substantial deficit. This major provider role has been assumed, at least in part, by the BRIIAT. As can be seen, in this region the trade balance with all regions improved over the period.

Conclusions

Most of the services traded are used as inputs. In the first part of this chapter we noted that the positive impact of the use of imported business services inputs has been extensively studied both in theoretical and empirical terms. Over the last decades the tradability of business services has greatly increased mainly due to advances in new technologies and falling in communication costs and to trade liberalisation. Moreover, in contrast to the 'traditional' location pattern that established a clear distinction between the types of activities that were produced in developed and developing countries, some emerging countries are acquiring a pre-eminent role in the provision of high-skilled services, such as is the case of business services. Thus, knowledge services clusters are emerging far from those countries classified as 'advanced' (Manning, 2013) and in the ever more and more global value chains, low-skilled work seems to show a declining value added share in emerging countries (Timmer, *et al.*, 2014). In addition, global cities have acquired a pre-eminent role as operational capacity providers (Sassen, 1991). All these 'new' trends give place to substantial changes in the map of world trade flows in business services inputs. The objective of this chapter was to analyse the evolution of these flows over the period 1995–2011 using a novel source: the WIOD. In particular we employed the regional aggregation of this database that classifies the 40 countries covered into six major regions: the Eurozone, the Other EU, the NAFTA, China, East Asia and the BRIIAT. The 'map' of world trade flows in business services inputs was 'drawn' in two phases: first, in addition to examine the evolution of the use of business services as inputs we analysed the changes in global trade flows in terms of trade balances and shares in world trade flows. Later, we explored the main changes in bilateral trade flows among regions.

The examination of the data referred to the use of business services inputs showed a general rise in the use of business services inputs, which, in some cases, like in East Asia and, especially, China, greatly relied on imports. In terms of the evolution of trade flows our empirical analysis supports the main location characteristics of the current wave of international sourcing of business services described by Massini and Miozzo (2012): the rapid relocation of business services to India and China that started in the 1990s has expanded to the rest of Asia, Eastern Europe and Latin and Central America in the 2000s. Thus, while the NAFTA experienced a drop in its share in world trade, China and the BRIIAT gained importance as global providers. Their growing role as exporters reflects in the radical change experienced by their trade balances from deficit to surplus. In the case of the Eurozone, the world leading exporter of business services inputs, it is necessary to highlight the 'exchange' of roles with its major trading partner, the Other EU countries. This can be explained, as Hardy *et al.* (2011) noted, because two main stages can be distinguished in foreign investment in business services in Central and Eastern Europe: over the 1990s only some oligopolistic firms established to provide business services non-available hitherto in these countries but since the 2000s specific business services activities are being relocated, most of them related to back office and corporate functions and customer care and technology support. In terms of bilateral trade flows, the concentration of trade among the EU and the NAFTA has also considerably diminished, thanks in part to the emergence of a new major client: China and an ever more important provider: the BRIIAT.

The changes in the global sourcing of business services inputs described above can serve as a starting point for future analyses. Our data on international trade in business services inputs comprises the international provision of these services, regardless whether the foreign provider is independent from the client firm or it is owned by the client firm. It would be very interesting to distinguish among the different modes of trade in service and to descend from the regional scale to the local scale because, as Taylor *et al.* (2013: 509) point out, 'descriptions of economic patterns have been equally territorialist and state-based'. For instance, if we descend to the city level in the analysis of producer services we can observe how although Chinese cities are increasingly important in terms of global network connectivity this is not reflected in strategicness (Taylor *et al.*, 2014). In addition, from a firm-level point of view, it is necessary to bear in mind that, as Breinlich and Criscuolo (2011) noted in their analysis for trade in services in the UK, there is a strong degree of heterogeneity among firms within the same industry.

References

Abramowsky, L. and Griffith, R. (2006). Outsourcing and offshoring of business services: How important is ICT?, *Journal of the European Economic Association*, 4(2–3), pp. 594–601.

Amiti, M. and Wei, S. (2005). Fear Of Service Outsourcing: Is it Justified? *Economic Policy,* 20(42), pp. 308–47.

Amiti, M. and Wei, S. (2009). Service Offshoring and Productivity: Evidence from the US. *The World Economy*, 32(2), pp. 203–20.

Arnold, J.M., Javorcik, B.S., Lipscomb, M. and Mattoo, A. (2012). Services Reform and Manufacturing Performance: Evidence from India. *Policy Research Working Paper*, 5948, Washington, DC: World Bank.

Arnold, J.M., Javorcik, B.S. and Mattoo, A. (2011). Does Services Liberalization Benefit Manufacturing Firms? Evidence from the Czech Republic. *Journal of International Economics*, 85(1), pp. 136–46.

Arora, A. and Gambardella, A. (2005). *From Underdogs to Tigers: The Rise and Growth of the Software Industry in Brazil, China, India, Ireland and Israel*. New York: Oxford University Press.

Backer, K.D. and Miroudot, S. (2013). Mapping Global Value Chains. *OECD Trade Policy Paper*, 159. Paris: OECD.

Baldwin, R.E. and López-González, J. (2013). Supply-Chain Trade: A Portrait of Global Patterns and Several Testable Hypotheses. *CEPR Discussion Paper*, 9421. London: CEPR.

Bhagwati, J.N. (1984). Splintering and disembodiment of services and developing nations. *The World Economy*, 7(2), pp. 133–43.

Blinder, A. (2006). Offshoring: the next industrial revolution? *Foreign Affairs*, 85, pp. 113–28.

Breinlich, H. and Criscuolo, C. (2011). International trade in services: A portrait of importers and exporters. *Journal of International Economics*, 84(2), pp. 188–206.

Chaminade, C. and Vang, J. (2008). Globalisation of knowledge production and regional innovation policy: Supporting specialized hubs in the Bangalore software industry. *Research Policy*, 37(10), pp. 1648–95.

Cheung, C., Rossiter, J. and Zheng, Y. (2008). Offshoring and Its Effects on the Labour Market and Productivity: A Survey of Recent Literature. *Bank of Canada Review*, Autumn, pp. 15–28.

Daniels, P.W. (2001). The locational geography of advanced producer services firms in the United Kingdom. *Progress in Planning*, 43, pp. 123–38.

Dunning, J.H. and Norman, G. (1983). The theory of multinational enterprise: an application of multinational office location. *Environment and Planning A*, 15, pp. 675–92.

Faulconbridge, J.R., Hall, S. and Beaverstock, J.V. (2008). New insights into the internationalization of producer services: Organizational strategies and spatial economies for global headhunting firms. *Environment and Planning A*, 40, pp. 210–234.

Florida, R. (2005). The World is Spiky. *The Atlantic Monthly,* 296, pp. 48–51.

Freund, C. and Weinhold, D. (2002). The Internet and International Trade in Services. *American Economic Review*, 92(2), pp. 236–40.

Friedman, T. (2005). *The World is Flat: A Brief History of the 21st Century*. London: Penguin.

Gereffi, G. and Fernandez-Stark, K. (2010). The offshore services value chain: developing countries and the crisis. In: O. Cattaneo, G. Gereffi and C. Staritz (eds), *Global Value Chains in a Postcrisis World. A Development Perspective*. Washington, DC: The World Bank, pp. 335–72.

Gertler, M.S. (1995). “Being there”: Proximity, Organization, and Culture in the Development and Adoption of Advanced Manufacturing Technologies. *Economic Geography*, 71(1), pp. 1–26.

Gertler, M.S. (2003). Tacit knowledge and the economic geography of context, or The undefinable tacitness of being (there). *Journal of Economic Geography*, 3(1), pp. 75–99.

Görg, H. and Hanley, A. (2005). International Outsourcing and Productivity: Evidence from the Irish Electronics Industry. *North American Journal of Economics and Finance*, 16(2), pp. 255–69.

Görg, H., Hanley A. and Strobl, E. (2008). Productivity Effects of International Outsourcing: Evidence from Plant Level Data. *Canadian Journal of Economics*, 41(2), pp. 670–688.

Goswami, A.G., Mattoo, A. and Sáez, S. (2011). *Exporting Services. A Developing Country Perspective*. Washington, DC: World Bank.

Guerrieri, P. and Meliciani, V. (2005). Technology and international competitiveness: The interdependence between manufacturing and producer services. *Structural Change and Economic Dynamics*, 16(4), pp. 489–502.

Hardy, J., Sass, M. and Fifekova, M.P. (2011). Impacts of horizontal and vertical foreign investment in business services: the experience of Hungary, Slovakia and the Czech Republic. *European Urban and Regional Studies*, 18(4), pp. 427–43.

Hijzen, A., Tomohiko, I. and Yasuyuki, T. (2010). Does Offshoring Pay? Firm-level Evidence from Japan. *Economic Enquiry*, 48(4), pp. 880–95.

Jensen, B. and Kletzer, L.G. (2005). Tradable services: Understanding the scope and impact of service outsourcing. Institute for International Economics Working Paper 05–9. Washington, DC: Institute for International Economics.

Jensen, J.B. (2013). Overlooked Opportunity: Tradable Business Services, Developing Asia, and Growth. *Asian Development Bank Working Paper*, 326. Philippines: Asian Development Bank.

Jensen, J., Rutherford, T.F. and Tarr, D.G. (2007). The Impact of Liberalizing Barriers to Foreign Direct Investment in Services: The Case of Russian Accession to the World Trade Organization. *Review of Development Economics*, 11, pp. 482–506.

Jensen, J.B. and Kletzer, L.G. (2010). Measuring Tradable Services and the Task Content of Offshorable Services Jobs. In K.G. Abraham, J.R. Spletzer and M. Harper (eds), *Labor in the New Economy*. Chicago: University of Chicago Press, pp. 309–35.

Johnson, R.C. and Noguera, G. (2012). Proximity and Production Fragmentation. *American Economic Review Papers and Proceedings*, 102(3), pp. 407–11.

Knobel, A. (2012). The Influence of Services Trade Liberalization on Services Flows and Industry Productivity in the CIS Countries and Russia. *Economics Education and Research Consortium Working Paper*, E12/05E. Available at: www.eerc.ru/paper-info/328 [accessed on 10 September 2015].

Los, B., Timmer, M.P. and de Vries, G.J. (2015). How Global are Value Chains? A New Approach to Measure International Fragmentation. *Journal of Regional Science*, 55(1), pp. 66–92.

Malecki, E.J. (2010). Global knowledge and creativity: new challenges for firms and regions. *Regional Studies*, 44(8), pp. 1033–52.

Maskell, P. and Malmberg, A. (1999). Localised learning and industrial competitiveness. *Cambridge Journal of Economics*, 23(2), pp. 167–86.

Manning, S. (2013). New Silicon Valleys or a new species? Commoditization of knowledge work and the rise of knowledge services clusters. *Research Policy*, 42(2), pp. 379–90.

Manning, S., Ricart, J.E., Rosatti Rique, M.S. and Lewin, A.Y. (2010). From blind spots to hotspots: how knowledge services clusters develop and attract foreign investment. *Journal of International Management*, 16(4), pp. 369–82.

Markusen, J. R. (1989). Trade in Producer Services and in Other Specialized Intermediate Inputs. *American Economic Review*, 79(1), pp. 85–95.

Markusen, J.R., Rutherford, T.R. and Tarr, D. (2005). Trade and direct investment in producer services and the domestic market for expertise. *Canadian Journal of Economics*, 38(3), pp. 758–77.

Massini, S. and Miozzo, M. (2012). Outsourcing and offshoring of KIBS: challenges to theory, management and geography of innovation. *Regional Studies*, 46(9), pp. 1219–42.

Michel, B. and Rycx, F. (2014). Productivity Gains and Spillovers from Offshoring. *Review of International Economics*, 22(1), pp. 73–85.

Miles, I., Kastrinos, N., Flanagan, K., Bilderbeek, R., Den Hertog, P., Huntink, W. and Bouman, M. (1995). Knowledge-intensive business services. European Innovation Monitoring Systems (EIMS) Publication No. 15. Innovation Programme, DGXIII, Luxembourg.

Mudambi, R. (2008). Location, control and innovation in knowledge-intensive industries. *Journal of Economic Geography*, 8(5), pp. 699–725.

OECD. (2013). *Interconnected Economies: Benefiting from Global Value Chains*. Paris: OECD.

Parthasarathy, B. and Aoyama, Y. (2006). From software services to R&D services: local entrepreneurship in the software industry in Bangalore, India. *Environment and Planning A*, 38(7), pp. 1269–85.

Sassen, S. (1991). *The Global City*. Princeton, NJ: Princeton University Press.

Taylor, P.J., Derudder, B., Faulconbridge, J., Hoyler, M. and Ni, P. (2014). Advanced Producer Service Firms as Strategic Networks, Global Cities as Strategic Places. *Economic Geography*, 90, pp. 267–91.

Taylor, P.J., Derudder, B., Hoyler, M. and Ni, P. (2013). New regional geographies of the world as practised by leading advance producer service firms in 2010. *Transactions of the Institute of British Geographers*, 38, pp. 497–511.

Timmer, M.P. (ed.) (2012). The World Input-Output Database (WIOD): Contents, Sources and Methods. *WIOD Working Paper*, 10. Groningen: University of Groningen.

Timmer, M.P., Erumban, A.A., Los, B., Stehrer, R. and de Vries, G.J. (2014) Slicing Up Global Value Chains. *Journal of Economic Perspectives*, 28(2), pp. 99–118.

Van der Marel, E. and Shepherd, B. (2013). Services Trade, Regulation and Regional Integration: Evidence from Sectoral Data. *The World Economy*, 36(11), pp. 1393–1405.

Winkler, D. (2010). Services Offshoring and Its Impact on Productivity and Employment: Evidence from Germany, 1995–2006. *The World Economy*, 33(12), pp. 1672–1701.

Wolfe, R. (2015). First Diagnose, Then Treat: What Ails the Doha Round?. *World Trade Review*, 14(1), pp 7–28.

Zhang, Y., Tang, Y. and Findlay, C. (2010). Productivity Effects of Services Trade Liberalization: Evidence from Chinese Firm-level Data. In: *International Conference on China and the Global Economy: Economic Integration and Protection*. Beijing, pp. 12–13, June 2010.

3 Services and growth in developing countries

A Kaldorian analysis

Gisela Di Meglio

Introduction

Research on structural change and economic development seemed to have reached a stagnant point after the pioneering works by Clark (1957), Kaldor (1966), Kuznets (1966), and Chenery *et al.* (1986). However, a recent set of studies has put the dynamics of the sectoral composition of productive structures of economies and its impact on growth back in the spotlight. Interest has been renewed across developed countries (Jorgenson and Timmer, 2011) and, particularly, across developing economies (Timmer and de Vries, 2009; Bah, 2011; McMillan and Rodrik, 2011; McMillan, *et al.*, 2014; Timmer *et al.*, 2014; Rodrik, 2015).[1]

The structure of an economy matters for growth and development (Thirlwall, 2013) since sectors have different capabilities to induce productivity gains, promote the expansion of other sectors or benefit from internal and external demand growth (Cimoli *et al.*, 2005). In the Kaldorian tradition, the manufacturing industry has been considered as the main vector of productivity growth and thus as the 'engine of growth' of economies. Kaldor (1966, 1967) argued that there is a need to distinguish between increasing returns activities, on the one hand (associated with manufacturing), and diminishing returns activities, on the other (associated with agriculture). Manufacturing expansion can induce productivity growth in other sectors by pulling labour away from them, which ultimately fosters overall productivity growth. Therefore, industrialisation is seen as a pathway to output growth and, ultimately, economic development.

These old ideas on the pivotal role played by manufacturing in growth are challenged by new evidence. First, 'premature deindustrialisation' is taking place in several developing economies as documented by Palma (2005); Dasgupta and Singh (2005, 2006) and Rodrik (2015). In these countries, the share of manufacturing in employment and value-added is shrinking at levels of income per capita that are much lower than those at which advanced nations began to deindustrialise. Thus, many developing countries are becoming service-based economies faster than developed countries did during their transition through development stages (Ghani and Kharas, 2010; Bah, 2011; Szirmai, 2012). Second, the tradability of services has progressively increased and value chains of services production have now globally expanded. This means that the advantages that manufacturing and agriculture traditionally had on the trade/globalisation sphere are now eroding (Dasgupta and Singh, 2005). The increasing interdependence of countries implies that an integrative and comparative view of structural change has become much more relevant (Matsuyama, 2009). Third, some service subsectors have completely broken away from old myths and have turned into some of the most innovative and productive activities. Productivity gains in market services are now as important as, or even more important than, productivity growth in manufacturing (Timmer and de Vries, 2007; Timmer and de Vries, 2009; Maroto-Sánchez and Cuadrado-Roura, 2009; United Nations, 2010). At the same time, after the financial crisis, the debate is going towards the 'reindustrialisation' of advanced countries and the fostering of industrial policies as a means to achieve sustainable growth paths (Tregenna, 2011; Westkämper, 2014).

Facing these old ideas with the new evidence, this chapter revisits the links between services, economic growth and development. Its main objective is to identify whether some specific service activities can act as an engine of economic growth. This is done by analysing structural change and sectoral productivity-growth patterns in developing countries. The underlying hypothesis is that the same productivity-growth connections that Kaldor uncovered for the manufacturing sector may apply to knowledge and technology based-services with strong inter-industry linkages. Whether services-led growth can become a new growth model for developing countries in the absence of sizeable manufacturing industries is a matter of concern (Rodrik, 2015). This timely investigation aims to contribute to this debate through a Kaldorian lens, one of the classical theoretical approaches in relation to the links between sectoral structure and productivity growth.

The chapter comprises four sections. Section 2 presents the concepts and theoretical framework developed by Kaldor. Section 3 discusses structural change in developing Asia, Latin America and Africa by means of fresh empirical data. Then, long-term productivity-growth connections at sectoral level are analysed in Section 4 using a sample of 27 developing countries between 1975 and 2005.The last section summarises the key results of this chapter and identifies future research paths.

Structural change and growth: the Kaldorian framework

According to Nicholas Kaldor it is impossible to understand economic growth and development without adopting a sectoral approach in which Lewis-type dualisms can be found (Lewis, 1954). Moreover, the supply and demand characteristics of each sector of the economy (agriculture, industry and services) needs to be taken into account. On the demand side, Kaldor (1966, 1967) supposes that the income elasticity of manufacturing is similar to that of services but higher than that of agriculture.[2] On the supply side, he suggests that a low productivity agricultural sector with a labour surplus, subject to diminishing returns, coexists with a capital-intensive manufacturing sector that is characterised by increasing returns and high productivity growth. At the same time, the productivity growth in services tends to be considerably lower than that of manufacturing. Manufacturing shows greater potential for productivity growth than other sectors due to the presence of static and dynamic economies of scale in production processes. Static economies of scale are mainly technical, internal to the firm, and related to mass production, whereas dynamic economies refer to increasing productivity derived from 'learning by doing' (Arrow, 1962) and from the so-called 'macroeconomic of scale' in the sense of Young (1928).[3]

Considering these conditions, the reallocation of labour from less productive activities (agriculture) to more productive sectors fosters productivity growth in both sectors and depends on the demand produced by manufacturing expansion. In the Kaldorian approach, manufacturing growth is not resource-constrained but rather demand-determined. In the first stages of development, the expansion of this key sector is closely fuelled by the demand of the dominant agricultural sector while in later stages it is mainly driven by external demand or export growth. A virtuous circle of cumulative growth can be explained on the basis of external demand evolution (Kaldor, 1970; Dixon and Thirlwall, 1975). As exports grow, output increases lead to fast productivity growth (by the presence of dynamic economies of scale) which induces an increase in competitiveness that leads to an additional increase in exports.

Kaldor's growth laws

In the view of these stylised facts, Kaldor articulated a set of long-run relationships between output, employment and productivity growth at the sectoral level of the economy, generally known as Kaldor's Growth Laws. According to these laws, there is a close association between the expansion of industrial activity in a country (particularly in manufacturing) and overall output growth via productivity. Therefore, manufacturing can be conceived as an 'engine of growth' (Thirlwall, 1983).

The first law states that the faster the growth of manufacturing output in an economy, the faster the growth of gross domestic product (GDP). Due to the production characteristics of manufacturing previously outlined, Kaldor defined this relationship as a fundamentally *causal* one. The second law states that the faster the expansion of manufacturing output, the faster will be the growth of productivity in manufacturing as a result of increasing returns to scale in this sector. This law is also known as *Verdoorn's Law*, after the Dutch economist P.J. Verdoorn who found such a relationship in the case of Eastern European countries in the 1940s. This second law can also be interpreted from the perspective of employment growth in manufacturing: the higher the economies of scale of the sector, the lower the employment elasticity with respect to output. Verdoorn's law itself has been subject to extensive debate and thus generated a large body of empirical and theoretical literature.[4] The third law states that the faster the growth of the manufacturing output, the faster the reallocation of labour from other sectors of the economy subject to diminishing returns to scale which raises productivity in both sectors. Productivity growth in manufacturing is explained by operation of the *Verdoorn's Law*. Moreover, as Kaldor supposes that in the first stages of development there is disguised unemployment in agriculture, the transfer of labour from this sector into industry does not reduce output and thus productivity in agriculture increases.

Kaldor confirmed that manufacturing was an engine of growth for twelve OECD countries using cross sectional data over the period 1953 to 1964. Since then, the interpretation and the validity of the different laws have been examined across developed countries (Rowthorn, 1975; Parikh, 1978; McCombie, 1983; Michl, 1985); for individual countries (Bairam, 1991; Atesoglu, 1993); across regions within countries (McCombie and de Ridder, 1984; León-Ledesma, 2000) and across regions across countries (Pons-Novell and Viladecans-Marsal, 1999). Recently, several studies have argued that manufacturing is the engine of growth in developing economies.[5] Felipe (1998) finds evidence across five Southeast Asian countries and Wells and Thirlwall (2003) across 45 African countries. In the case of Latin American economies, evidence of increasing returns to scale in industrial sectors is found by Cimoli *et al.* (2005) and Libano (2006). Pacheco and Thirlwall (2014) consider the role played by export growth in open developing economies and conclude that there is a strong association between export and economic growth.

In contrast, very few studies have examined whether services may serve as a means for growth in developing economies. Dasgupta and Singh (2005; 2006) are the first ones to argue, within a Kaldorian framework, that although manufacturing continues to be critical for development, services can also be regarded as an additional engine of growth. In particular, they emphasise the role played by ICT related services in the context of India. Services also have productivity growth-inducing effects through the exploitations of scale economies in developing Asia –although to a lesser extent than manufacturing– as discussed by Felipe *et al.* (2009). Notwithstanding the productive heterogeneity of the region, they emphasise that the composition of services matters for determining the sustainability of their contribution to productivity growth. In fact, most studies fail to deal with the heterogeneity of the services sector in developing economies. A notable exception is Pieper (2003), who finds strong evidence for increasing returns to scale at sectoral level across 30 developing countries. This finding is relevant since it suggests that the same Kaldorian mechanisms that make manufacturing the engine of growth may also apply to service activities.

Nevertheless, a couple of caveats should be added to the aforementioned studies. The Kaldorian framework relies on an unproblematic productivity measurement of manufacturing, whereas productivity in services faces conceptual and methodological issues that are far from being resolved. This is particularly relevant in case of non-market services (Di Meglio *et al.*, 2015). The measurement of productivity in public services is in a pre-paradigmatic phase in which efforts to improve the measurement of output (and outcomes) in services are being made and different approaches are being tested. Hence, country comparisons in time and space are often risky and any empirical analysis should be interpreted with caution. Moreover, at the time when the Kaldorian framework was developed, it was easier to make a clear-cut classification of sectors in an economy. At present, these distinctions are somewhat artificial due to the blurring of sectoral boundaries emerging from the increased reciprocal relationships between services and manufacturing. Notwithstanding these caveats, the concepts developed by Kaldor constitute a useful framework to analyse changes in the composition of production and employment, and its interaction with patterns of sectoral productivity growth. The following sections of this chapter empirically investigate these topics for the case of developing economies.[6]

Paths of structural transformation in the developing world

In general, developed countries have all followed the same well-known process of structural transformation: a declining share of agriculture in output as GDP per capita increases, then an inverted U-shape relationship between industry shares and income per capita followed by an increasing share of services. However, developing countries nowadays show a distinguishable feature: a very large share of services in output at early stages of development (Szirmai, 2012). This is explained by the fact that, since 1990, the hump-shaped relationship between industrialisation and income per capita has shifted downwards and moved closer to the origin (Rodrik, 2015). This means that developing countries are deindustrialising faster and at much lower levels of income than early industrialisers when they reached their industrial peak production. Heterogeneity critically matters since there is not one single process of structural change either across developing regions or within each of them. Bah (2011) argues that the pattern of structural change followed by Asia is the closest to that of the developed countries, while African and Latin American countries have comparatively higher service output shares at very low levels of income per capita. Rodrik (2015) also emphasises that both low-income countries in Sub-Saharan Africa and middle-income economies in Latin America have been strongly hit by deindustrialisation, while Asian countries (and industrial exporters) have been insulated from this trend.

Sectoral growth patterns in the developing world are shown in Figure 3.1, in which employment and output shares are plot against (the log of) income per capita for a pool of 29 countries during the period 1950–2010. Data on sectoral shares comes from the Groningen Growth and Development Centre 10-Sector Database (Timmer and de Vries, 2007) and the Africa Sector Database (de Vries *et al.*, 2013). The country sample includes nine countries from Latin America, nine from Asia and eleven from Sub-Saharan Africa (for more details, see the Appendix). Data for constructing income per capita (real GDP at constant prices in mil. 2005 dollars and population) comes from the Penn World Table 8.0 (Feenstra *et al.*, 2013). The first panel of Figure 3.1 confirms the steady decline in agricultural employment and output shares as income increased during the past five decades. The heterogeneity between regions and countries is also clear from the graphs. For example, on average, African countries have higher agricultural

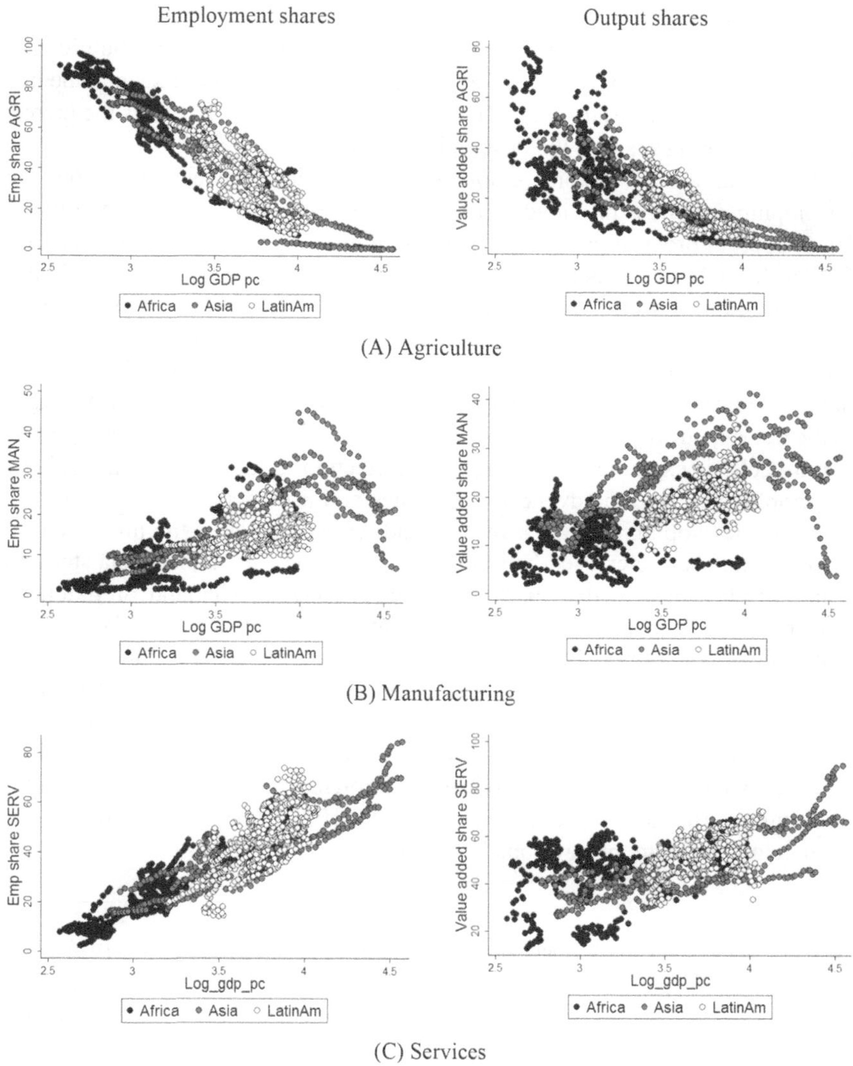

Figure 3.1 Structural transformation in developing economies, 1950–2010

Note: Data on value added at current prices is used for constructing output shares. Each data point indicates a country's sectoral share in a particular year. Agriculture refers to sectors A and B of ISIC Rev.3.1, manufacturing to sector D and services comprise sectors G to Q. See Appendix for details on the country coverage and data availability.

Source: Based on GGDC 10-Sector Database and the Africa Sector Database and PWT 8.0.

shares compared to Asian or Latin American countries. However, in some Asian countries (like Thailand or India) this sector remains as important as in many African economies. The inverted U-shape between manufacturing and income per capita traditionally followed by advanced countries is more evident for the

Asian countries in our sample, among which we may find the relatively most industrialised countries (Malaysia and Taiwan). In contrast, poorer Sub-Saharan African economies and middle-income Latin American countries seem to be in an earlier phase of industrialisation. The last panel of Figure 3.1 reveals the increasing role of services across the developing world.

A comparison of sectoral employment allocation between developed and developing regions is provided in Table 3.1. The time span has been confined to 1975–2005 in order to provide comparable information across regions. The first finding from Table 3.1 is that the share of agricultural employment is still larger in the developing world than in advanced countries, despite its relatively decline in recent decades. This particularly holds for the African region, as previously observed. At the beginning of the period under analysis, the share of manufacturing employment in developing economies was, on average, nearly half the figure in the developed world (13.9 per cent versus 24.7 per cent). The narrowing of this gap by 2005 is explained by a long-run contraction in the shares of manufacturing employment in the advanced economies and a moderate contraction in the shares of developing countries. In Asia and Africa, however, the manufacturing employment share increased between 1975 and 1985. Since then Asia started to deindustrialise despite that many countries (Malaysia, Thailand or Indonesia) have raised their manufacturing employment shares. In this region, labour input reallocated to services, but (declining) manufacturing shares have remained comparatively higher compared to other developed and developing regions. In Africa, manufacturing shares have remained rather stable at around 8 per cent during 1985–2005, a figure that is half of the Asian shares in 2005. It is in Latin America where the process of deindustrialisation has been more intense: the share of manufacturing steadily declined during the 1975–2005 period and it is now even below the share of developed countries.

Table 3.1 Patterns of employment shares: developed vs developing regions, 1975–2005

	1975			*1985*			*1995*			*2005*		
	AGR	*MAN*	*SERV*	*AGR*	*MAN*	*SERV*	*AGR*	*MAN*	*SERV*	*AGR*	*MAN*	*SERV*
Latin America	33.9	15.5	41.9	28.6	14.5	49.0	23.1	14.6	53.9	18.4	12.5	60.9
Asia	42.8	18.3	33.8	34.7	19.1	39.6	26.1	18.6	47.2	22.7	16.5	53.1
Africa	66.0	7.8	20.9	64.6	8.2	22.6	60.5	8.2	26.3	54.0	8.2	32.9
Developing countries	47.6	13.9	32.2	42.6	13.9	37.1	36.6	13.8	42.5	31.7	12.4	49.0
Developed countries [(a)]	9.2	24.7	56.4	6.8	20.8	64.2	4.4	17.9	70.2	3.1	14.5	74.3

Notes: (a) United States, Denmark, Spain, France, United Kingdom, Italy, the Netherlands, and Sweden. Unweighted regional averages. AGR stands for agriculture, MAN for manufacturing and SERV for services. See Appendix for details on the country coverage and data availability.

Source: Based on GGDC 10-sector database and the Africa sector database.

The structural transformation of developing countries has followed, to a certain extent, uneven patterns across regions. It should be pointed out that most advanced countries have experienced employment deindustrialisation rather than output deindustrialisation suggesting a particular type of unskilled labour-saving type of technological progress (Rodrik, 2015). However, for developing countries, deindustrialisation is related to how economies manage globalisation and, in particular, the decline in the relative price of manufactures on the world market (Rodrik, 2015). A strong reduction in both industrial employment and output shares is generally observed, with the exception of economies with a strong comparative advantage in manufacturing. In particular, most Asian exporters' countries have managed to avoid the shrinking of manufacturing output share and the decline of manufacturing employment shares is less pronounced than in the rest of the developing world.

Sectoral engines of productivity growth in developing regions

Methodological issues

Diverse patterns of structural change may contribute differently to productivity growth and, thus, to economic performance across developing regions. Therefore, this section studies sectoral contributions to labour productivity growth in Asia, Latin America and Africa between 1975 and 2005. A preferable measure of productivity would be Total Factor Productivity (TFP).[7] However, the lack of data availability for physical and human capital at sectoral level prevents us to use it in this section.

Following the procedures outlined by OECD (2001), the value-added based measure of labour productivity growth by sector (π_j) is given by the relation: $\pi_j = \hat{q}_j - \hat{l}_j$ where $\hat{q}_j$ denotes the rate of change of value added at constant prices in sector j and $\hat{l}_j$ the rate of change of labour input in sector j. The aggregate rate of change in value added is a share-weighted average of the sector-specific rate of value added change, where weights reflect the share of each sector in value added at current prices:

$$\hat{q} = \sum_j s_j^q \cdot \hat{q}_j \text{ where } s_j^q = p_j\, q_j / \sum_j p_j\, q_j$$

On the input side, aggregation of sector-level labour input is achieved by weighting the growth rates of employees by sector with each industry's share in total employment:

$$\hat{l} = \sum_j s_j^l \cdot \hat{l}_j \text{ where } s_j^l = l_j / \sum_j l_j$$

Aggregate labour productivity growth (π) is defined as the difference between aggregate growth in value-added and aggregate growth in labour input:

$$\pi = \sum_j (s_j^q \cdot \hat{q}_j - s_j^l \cdot \hat{l}_j)$$

The contribution of a particular sector to aggregate labour productivity growth is $(s_j^q \cdot \hat{q}_j - s_j^l \cdot \hat{l}_j)$, or the difference between its contribution to total value added and to total labour input.

Sectoral contributions to growth in developing countries

Following the aforementioned methodology, Figure 3.2 shows each industry's contribution to productivity, value added and employment growth between 1975 and 2005, while Figure 3.3 details the contribution of the service activities. Employment has expanded at a similar pace across developing economies during the past three decades (2.6 per cent per year on average) with services playing a significant role in inter-sectoral employment shifts.[8] This sector accounts for around three-quarters of job expansion in both Asia and Latin America, a contribution that exceeds the participation of services in total employment in 2005 (see Table 3.1). As expected, the impact of services is somewhat smaller than that of agriculture in Africa, although together they can explain almost 90 per cent of job creation. Manufacturing accounts for 14 per cent of employment growth in Asia, 10 per cent in Latin America and 8 per cent in Africa. All these figures are quite close to the share of this sector in total employment in 2005 (see Table 3.1).[9] Figure 3.3 allows us to understand growth patterns within service activities. Commerce (comprising wholesale and retail trade and hotels and restaurants) accounts for the most important part of services employment change in every region under study followed by public services; business services and transport and communications.

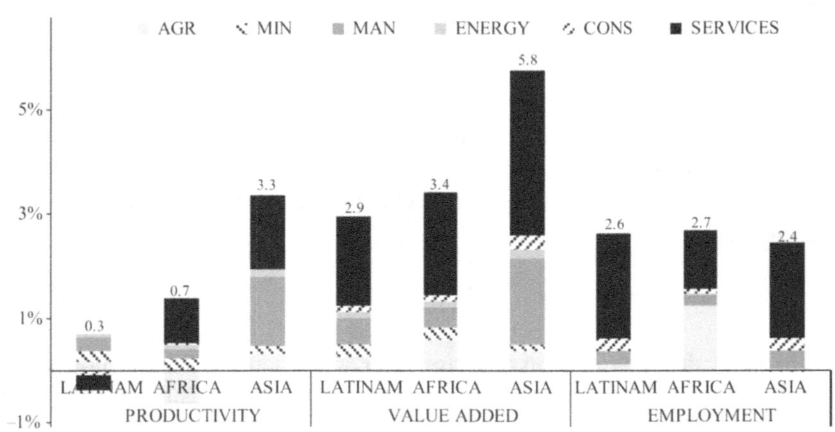

Figure 3.2 Sectoral contributions to growth 1975–2005, by regions

Notes: Unweighted averages. AGR: agriculture (sectors A+B of ISIC Rev.3.1), MIN: mining and quarrying (sector C), MAN: manufacturing (sector D), ENERGY: public utilities (sector E), CONS: construction (sector F), SERVICES: sectors G to Q of ISIC Rev.3.1. Latin America (LATINAM) excludes Brasil and Peru due to data availability. Bolivia and Venezuela 2003 instead of 2005; India 2004. See Appendix for more details.

Source: Based on GGDC 10-Sector Database and the Africa Sector Database.

Since labour input has expanded at a quite similar rate, patterns in productivity growth across developing regions are explained by their diverse speed in value added creation between 1975 and 2005. Asia shows the highest growth rate of output (5.8 per cent per year), twice the figure for Latin America, while in Africa value creation expands at 3.4 per cent per year.[10] Even though services explain more than half of this growth in every region; their contribution to value added is considerably less than in employment creation (except for Africa). In contrast, manufacturing behaves rather differently across developing areas and contributes more to value added growth than to employment growth. Manufacturing accounts for one-third of Asia´s output growth, while it explains 17 per cent of value added change in Latin America and 11 per cent in the case of Sub-Saharan Africa. According to Figure 3.3, patterns of value added growth within service activities also differ across regions. Public services are the most relevant category in both Latin America and Africa, while the rest of the sectors have contributed quite similarly to services output change (with the exception of transport in Africa). In contrast, value added growth in Asian services is mainly explained by commerce (accounting for 35 per cent), followed by public services and business services (explaining around 24 per cent each) and transport (17 per cent).

From the previous analysis, it is clear that Asia has experienced a much more rapid productivity increase between 1975 and 2005 than Africa or Latin America. Productivity growth in Latin America is hindered by the negative contribution of services suggesting adverse dynamic implications of structural change that has continuously reallocated workers to this sector after 1975.[11] Transport and

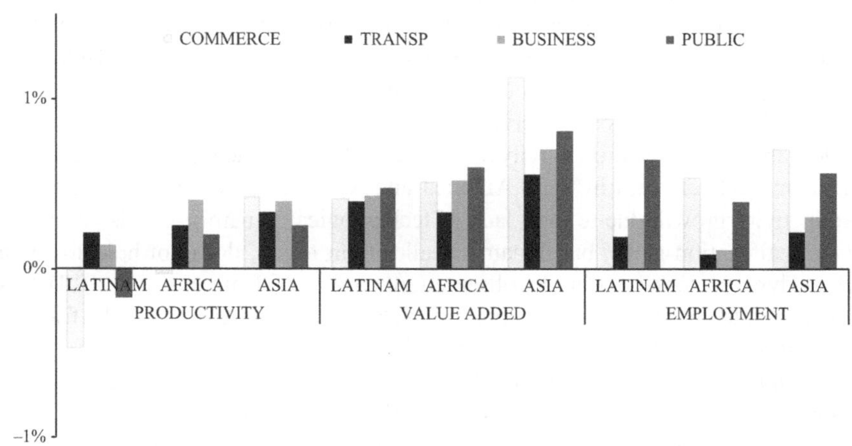

Figure 3.3 The contribution of services to growth 1975–2005, by regions

Notes: Unweighted averages. COMMERCE: wholesale and retail trade and hotels and restaurants (sectors G+H of of ISIC Rev.3.1); TRANSP: transport, storage and communications (sector I); BUSINESS: financial intermediation, real estate, renting and business activities (sectors J+K); PUBLIC: community, social, personal and government services (sectors L to P). Latin America (LATINAM) excludes Brasil and Peru due to data availability. Bolivia and Venezuela 2003 instead of 2005; India 2004. See Appendix for more details.

Source: Based on GGDC 10-Sector Database and the Africa Sector Database.

business services could not compensate for the bad performance of commerce and public services in terms of output per employee. It is manufacturing (together with mining) that explains the bulk of the slight productivity increase in Latin America. In contrast, services contribute more than manufacturing to the modest productivity growth experienced by Africa.[12] The positive impact of business activities and transport counterbalance the weak productivity performance in commerce. Here, agriculture contributes negatively to output per worker indicating potential gains from further inter-sectoral labour shifts. In Asia, fast productivity growth is equally enhanced by manufacturing and services.[13] Commerce accounts for a significant part of the services contribution to output per worker growth followed by business activities, transport and public services. Positive sectoral contributions point to positive dynamic impacts of structural change in Asia between 1975 and 2005.

These results are in line with recent literature focusing on structural change and growth in developing countries. By means of shift-share techniques, McMillan and Rodrik (2011) find that Asian countries have experienced growth-enhancing structural change between 1990 and 2005 while in Africa and Latin America growth-reducing structural change has prevailed. This indicates that labour has moved from high productivity sectors (like manufacturing or even some tradable services) to less productive activities (like personal services, informality or even unemployment).[14] An updated study shows that after 2000 structural change has been growth-enhancing in some African countries due to small expansions in the manufacturing sector (McMillan *et al.*, 2014). According to Timmer and de Vries (2009), market services (namely, distributive trades, financial services and transport and communications) seem to be the most important contributor to growth in developing Asia and Latin America from 1950 to 2005. Nevertheless, Timmer *et al.* (2014) find that when the country sample and the time horizon analysed are expanded, the role played by services in growth considerably differs. During the 1990s, above-average productivity *level* sectors that expanded in terms of employment (market services) in Latin America and Africa experienced below-average productivity growth due to their lack of technological dynamism. This pattern of static reallocation gains, but dynamic reallocation losses, does not hold for Asia where dynamic losses are hardly observed. According to Timmer *et al.* (2014), this may be related to the falling manufacturing shares in Latin America and Africa, in contrast with the stable shares showed by Asia.

All in all, productivity and growth seems to be related to the maintenance of a manufacturing sector capable of providing the base for the development of various producer-oriented services (such as logistics, R&D, finance, and so on.) which are more productive and prone to productivity increase rather than consumer-oriented retail and public services.

Concluding remarks and future research avenues

This investigation adopted a Kaldorian framework for understanding long-term growth in developing economies. This means that productivity growth connections in such regions are analysed from a sectoral perspective. The chapter finds diverse paths of sectoral transformation between the developed and the developing world in which a rapid emergence of services is taking place along with a declining contribution of industry. Further, diversity is evident, to some extent, between developing areas. Many Asian countries have already transitioned the inverted U-shape relationship between manufacturing and income per capita while Latin American and African economies seem to be in the first stage of industrialisation.

The chapter also finds that the reallocation of workers to service activities may enhance or reduce productivity and growth in developing economies. In the case of Asia, several interrelated factors explain the positive contribution of sectoral transformation to long-term growth. First, high value added creation in both manufacturing and services denote the export-oriented productive specialisation of a region deeply involved in global value chains. Moreover, not only the decline in Asian manufacturing shares has been smaller compared to other developing areas, but the structure of manufacturing production has technologically upgraded and has become more diversified, especially compared to the 1970s (Felipe and Estrada, 2007). Asian manufacturing has experienced a shift towards more skill-intensive sectors with higher productivity levels and growth during the past four decades (Jaumotte and Spatafora, 2007). The ADB (2007) also argues that in countries like Korea, Malaysia, Taiwan and Singapore manufacturing output is concentrated in more technology- and scale-intensive subsectors. Specialisation in high-tech export-oriented manufacturing with strong inter-industry linkages seems to allow for the development of dynamic service sectors (such as business-related service activities or highly tradable services) which stimulate productivity and growth.[15]

The situation in the rest of developing regions is rather different. Deindustrialisation in Africa is characterised by a declining diversity and sophistication of the region's manufacturing sectors (Page, 2011). In most Latin American economies, manufacturing is closely connected to natural resources intensive sectors (tobacco, coal, petrol, paper) (Cimoli *et al.*, 2005). This region, strongly hit by deindustrialisation, shows the worst regional productivity performance during 1975–2005.

All in all, Kaldor's old ideas remain valid to date: manufacturing has driven productivity and growth in developing regions. Moreover, the maintenance of a diversified, sophisticated, export-oriented manufacturing base has allowed for the development of producer-oriented services that drive growth as they are more productive and more prone to productivity increases compared to consumer-oriented services. This is particularly the case in Asia where certain services (mainly producer and advanced business services) have acted as an engine of growth. However, the profound deindustrialisation experienced by Latin America has prevented the development of many producer-oriented services. Instead, low-productive services combined with high informality levels have hampered growth. Between these two cases, we find Africa in which some tertiary activities (notably business services) have managed to develop and play a positive role in productivity growth.

Some scholars argue that the consolidation and stability of democracies may be hampered by premature deindustrialisation (Rodrik, 2015). In any case, the political consequences of the surge of services in developing areas are complex and deserve further scrutiny. Other issues still need to be addressed in the context of this investigation and offer a promising future research path. First, the econometrical estimation of Kaldor's growth laws would serve to quantitatively evaluate the engine of growth hypothesis. Estimations should be carried out by sector panels in order to take into account the heterogeneity of services and across developing areas. Moreover, an input-output analysis would allow for a better understanding of the relative contribution of final and intermediate demand to changes in the sectoral structure of developing economies. Finally, it would be interesting to take into account the heterogeneity of the manufacturing sector itself as well of producer-oriented services in order to have further insights in specific sectoral engines of growth in developing economies. In this respect, there is also scope for a more detailed analysis of which types of manufacturing activities are conducive to the growth of producer-oriented services, and, on the other hand, which types of export-oriented services apparently can flourish irrespective of the presence of a manufacturing base.

Acknowledgements

The author gratefully acknowledges funding through the Project: 'Services in developing economies: a new engine of growth?', XII Social Sciences National Call of the Ramón Areces Foundation, 2013–2015.

Notes

1 This new set of studies is closely related to the emergence and improvement of long-term series on sectoral output and productivity across the developing world.
2 However, the income elasticity of demand for services would tend to be higher than that of manufacturing products beyond a certain level of income per capita. This effect may, to some extent, be offset by the different price movements taking place in these two sectors, which have an influence on the rate of growth of demand. In particular, the lower relative prices for manufactures derived by faster productivity growth in this sector may also induce an increase in the demand for manufacturing products.
3 They reflect the macroeconomic spillover effects of the expansion of manufacturing activities (induced technological change, externalities in the production, etc.).
4 For a comprehensive review see McCombie *et al.* (2002).
5 At regional level, the pivotal role of manufacturing for the case of Chinese provinces is confirmed by Hansen and Zhang (1996) and Guo *et al.* (2012).
6 It is not the aim of this chapter to test Kaldor's growth laws in developing countries but rather to analyse their diverse process of structural change and of sectoral productivity patterns.
7 Productivity is defined as the ratio of output (or value added) to input(s). The two most commonly used measures are partial factor productivity and total factor productivity (TFP). In the former, the output is related to a single input and, in the latter the output is related to a combination of inputs. TFP is generally considered a measure of overall efficiency of labour and capital.

8 Of course, there are large variations across countries (for more details, see the Appendix). For example, employment has grown at 3.6 per cent in Kenya and Senegal and at 1.4 per cent in Argentina. Services contribution to employment growth also varies to a great extent reaching 3 per cent in Hong Kong and 0.4 per cent in Ethiopia.
9 In Malaysia manufacturing explains 33 per cent of employment growth whereas Argentina, Nigeria and Hong Kong saw declining manufacturing employment during the period under analysis (for more details, see the Appendix).
10 Again, differences between countries arise (for more details, see the Appendix).
11 With the exception of Chile and Mexico (for more details, see the Appendix).
12 With the exception of Nigeria and South Africa (for more details, see the Appendix).
13 With the exception of India and Korea (for more details, see the Appendix).
14 The authors argue that growth reducing structural transformation is related to the presence of large endowments of natural resources, the overvaluation of currencies and the reduced flexibility of labour markets. Pagés (2010) also argues that high levels of informality may explain the disappointing performance of service productivity in Latin America since inefficient firms are protected from competing with productive and efficient business models.
15 Many modern tradable services, like information and communication technology enabled services (call centres, finance, logistics and accounting, software development), have notably expanded in Asian countries with a relatively modest manufacturing base (India and the Philippines). Educated human resources at low cost, attractive fiscal incentives, entrepreneurial talent, infrastructure and industrial parks have driven the growth of these service industries (see Mitra, 2013).

References

ADB (2007). *Asian Development Outlook 2007*. Manila: Asian Development Bank.

Arrow, K. (1962). The Economic Implications of Learning by Doing. *Review of Economic Studies*, 29, pp. 155–73.

Atesoglu, H.S. (1993). Manufacturing and Economic Growth in the United States. *Applied Economics*, 25, pp. 67–9.

Bah, E. (2011). Structural Transformation Paths across Countries. *Emerging Markets Finance and Trade*, 47(2), pp. 5–19.

Bairam, E. (1991). Economic growth and Kaldor's law: the case of Turkey, 1925–78. *Applied Economics*, 23, pp. 1277–80.

Chenery, H.B., Robinson, S. and Syrquin, M. (1986). *Industrialization and Growth: A Comparative Study*. Oxford University Press: New York, USA.

Cimoli, M., Porcile, G., Primmi, A. and Vergara, S. (2005). Cambio Estructural, Heterogeneidad Productiva y Tecnología en América Latina. In: Cimoli, M. (ed.), *Heterogeneidad estructural, asimetrías tecnológicas y crecimiento en América Latina,* ECLAC, United Nations, available at: http://mpra.ub.uni-muenchen.de/3832/.

Clark, C. (1957). *The Conditions of Economic Progress*, 3rd ed. London: MacMillan.

Dasgupta, S. and Singh, A. (2005). Will Services Be the New Engine of Indian Economic Growth? *Development and Change*, 36(6), pp. 1035–58.

Dasgupta, S. and Singh, A. (2006). Manufacturing, Services and Premature De-industrialization in developing countries: a Kaldorian empirical analysis. CBR Working paper 327, June, University of Cambridge.

De Vries, G.J., Timmer, M.P. and de Vries, K. (2013). Structural Transformation in Africa: Static gains, dynamic losses. GGDC research memorandum 136, October, University of Groningen, Groningen Growth and Development Centre.

Di Meglio, G., Stare, M., Maroto, A. and Rubalcaba, L. (2015). Public services performance: an extended framework and empirical assessment across the enlarged EU. *Environment and Planning C: Government and Policy*, 33(2), pp. 321–41.

Dixon, R. and Thirlwall, A.P. (1975). A model of regional growth rate differences on Kaldorian lines. *Oxford Economic Papers*, 27, pp. 201–14.

Felipe, J. (1998). The Role of the Manufacturing Sector in Southeast Asian Development: A Test of Kaldor's First Law. *Journal of Post Keynesian Economics*, 20(3), pp. 463–85.

Felipe, J. and Estrada, G. (2007). Benchmarking Developing Asia's Manufacturing Sector. Asian Development Bank. Available at: https://openaccess.adb.org/handle/11540/1862#sthash.IDmuVCvB.dpuf [accessed April 2015].

Felipe, L., León-Ledesma, M., Lanzafame, M. and Estrada, G. (2009). Sectoral Engines of Growth in Developing Asia: Stylised Facts and Implications. *Malaysian Journal of Economic Studies*, 46(2), pp. 107–33.

Feenstra, R.C., Inklaar, R. and Timmer, M.P. (2013). The Next Generation of the Penn World Table. Available for download at www.ggdc.net/pwt.

Ghani, E. and Kharas, H. (2010). The service revolution. *Economic Premise*, 14, May, The World Bank.

Guo, D., Dall'erba, S. and Le Gallo, J. (2012). The Leading Role of Manufacturing in China's Regional Economic Growth: A Spatial Econometric Approach of Kaldor's Laws. *International Regional Science Review*, 36(2), pp. 139–66.

Hansen, J. and Zang, J. (1996). A Kaldorian approach to regional economic growth in China. *Applied Economics*, 28, pp. 679–85.

Jaumotte, F. and Spatafora, N. (2007). Asia Rising: A Sectoral Perspective. IMF Working Paper, WP/07/130.

Jorgenson, D.W. and Timmer, M.P. (2011). Structural Change in Advanced Nations: A New Set of Stylised Facts. *Scandinavian Journal of Economics*, 113(1), pp. 1–29.

Kaldor, N. (1966). *Causes of the Slow Rate of Economic Growth of the United Kingdom.* Cambridge: Cambridge University Press.

Kaldor, N. (1967). *Strategic Factors in Economic Development.* Ithaca NY: New York State School of Industrial and Labour Relations, Cornell University.

Kaldor, N. (1970). The Case for Regional Policies. *Scottish Journal of Political Economy,* 17, pp. 337–48.

Kuznets, S. (1966). *Modern Economic Growth: Rate, Structure and Spread*. London: Yale University Press.

León-Ledesma, M. (2000). Economic Growth and Verdoorn's Law in the Spanish Regions, 1962–91. *International Review of Applied Economics*, 14(1), pp. 55–69.

Lewis, W.A. (1954). Economic Development with Unlimited Supplies of Labour. *Manchester School of Economics and Social Studies*, 22, pp. 139–91.

Libanio, G. (2006). Manufacturing Industry and Economic Growth in Latin America: A Kaldorian Approach. Available at: www.policyinnovations.org/ideas/policy_library/data/01384/_res/id=sa_File1/Libanio_manufacturing.pdf [accessed December 2014].

Maroto-Sánchez, A. and Cuadrado-Roura, J.R. (2009). Is growth of services an obstacle to productivity growth? A comparative analysis. *Structural Change and Economic Dynamics*, 20(4), pp. 254–65.

Matsuyama, K. (2009). Structural Change in an Interdependent World: A Global View of Manufacturing Decline. *Journal of the European Economic Association*, 7 (2–3), pp. 478–86.

McCombie, J.S.L. (1983). Kaldor's Laws in Retrospect. *Journal of Post Keynesian Economics*, 5(3), 414–30.

McCombie, J. and de Ridder J. (1984). The Verdoorn Law Controversy: Some New Empirical Evidence Using US State Data. *Oxford Economic Papers,* 36, pp. 268–84.

McCombie, J., Pugno, M. and Soro, B. (2002). *Productivity Growth and Economic Performance: Essays on Verdoorn's Law.* London: Palgrave-Macmillan.

McMillan, M. and Rodrik, D. (2011). Globalization, Structural Change, and Productivity Growth. NBER working paper 17143.

McMillan, M.S., Rodrik, D. and Verduzco-Gallo, I. (2014). Globalization, structural change and productivity growth: with an update on Africa. *World Development*, 63, pp. 11–32.

Michl, T. (1985). International Comparisons of Productivity Growth: Verdoorn's Law Revisited. *Journal of Post Keynesian Economics*, 7, pp. 474–92.

Mitra, R. (2013). The Information Technology and Business Process Outsourcing Industry: Diversity and Challenges in Asia. ADB Economics Working Paper Series No. 365, September.

OECD (2001). *OECD Productivity Manual: A Guide to the Measurement of Industry-Level and Aggregate Productivity Growth.* Paris: OECD Publishing.

Pacheco, P. and Thirlwall, A.P. (2014). A new interpretation of Kaldor's first growth law for open developing economies. *Review of Keynesian Economics,* 3, pp. 384–98.

Page, J. (2011). Should Africa Industrialize? Working Paper Series UNU-WIDER Research Paper 2011/47, World Institute for Development Economic Research, UNU-WIDER.

Pagés, C. (2010). *The Age of Productivity: Transforming Economies from the Bottom Up.* Washington, USA: Inter-American Development Bank.

Palma, G. (2005). Four Sources of "De-industrialization" and a New Concept of Dutch Disease. In: Ocampo, J. (ed.), *Beyond Reforms: Structural Dynamics and Macroeconomic Vulnerability*. Palo Alto, CA: Stanford University Press and the World Bank.

Parikh, A. (1978). Difference in Growth Rates and Kaldor's Law. *Economica*, 45, pp. 83–91.

Pieper, U. (2003). Sectoral Regularities of Productivity Growth in Developing Countries: A Kaldorian Interpretation. *Cambridge Journal of Economics,* 27(6), pp. 831–50.

Pons-Novell, J. and Viladecans-Marsal, E. (1999). Kaldor's Laws and Spatial Dependence: Evidence for the European Regions. *Regional Studies*, 33, pp. 47–65.

Rodrik, D. (2015). Premature deindustrialization. NBER Working Paper 20935. Available at: www.nber.org/papers/w20935 [accessed on 30-05-2016].

Rowthorn, R.E. (1975). What remains of Kaldor's Law? *Economic Journal*, 85, pp. 10–19.

Szirmai, A. (2012). Industrialisation as an engine of growth in developing countries, 1950–2005. *Structural Change and Economic Dynamics*, 23(4), pp. 406–420.

Thirlwall, A.P. (1983). Symposium on Kaldor's Laws. *Journal of Post Keynesian Economics*, 5(3).

Thirlwall, A.P. (2013). *Economic Growth in an Open Developing Economy: The Role of Structure and Demand*, Cheltenham: Edward Elgar.

Timmer, M. and de Vries, G. (2007). A Cross-country Database For Sectoral Employment and Productivity in Asia and Latin America, 1950–2005. GGDC research memorandum 98, University of Groningen, Groningen Growth and Development Centre.

Timmer, M. and de Vries, G. (2009). Structural Change and Growth Accelerations in Asia and Latin America: A New Sectoral Data Set. *Cliometrica*, 3(2), pp. 165–90.

Timmer, M., de Vries, G. and de Vries, K. (2014). Patterns of Structural Change in

Developing Countries. GGDC research memorandum 149, University of Groningen, Groningen Growth and Development Centre.
Tregenna, F. (2011). Manufacturing Productivity, Deindustrialization, and Reindustrialization. Working Paper Series UNU-WIDER Research Paper 2011/57, World Institute for Development Economic Research,UNU-WIDER.
United Nations (2010). Structural Change and Employment Creation in Developing Countries. In: UN, *Trade and Development report*, New York and Geneva, pp. 99–131.
Wells, H. and Thirlwall, A.P. (2003). Testing Kaldor's Growth Laws across the Countries of Africa, *African Development Review*, 15, pp. 89–105.
Westkämper, E. (2014). *Towards the Re-Industrialization of Europe: A Concept for Manufacturing for 2030.* Springer.
Young, A. (1928). Increasing Returns and Economic Progress. *The Economic Journal,* 38(152), pp. 527–42.

4 Online services

An equalising force between the Global North and the Global South?

Robert Wentrup, H. Richard Nakamura and Patrik Ström

Introduction

During an Internet conference in Paris in December 2014 (LeWeb, 2014), Sir Tim Berners-Lee, often called the founder of the World Wide Web, was asked about its future. In an eloquent way he implied that the Internet has not yet come far in connecting people from different geographical regions, from different social groups, and from different ethnic origins or religions with each other. Seemingly, geographical, economic and cultural distances still prevail in the online realm, and people continue to communicate within limited geographical and socio-cultural spaces. Do online services have an equalising role in changing this pattern or do they fortify existing disparities?

The hyped concepts of 'death of the geography' (Cairncross, 2001) or a 'flat world' (Friedman, 2005) are antithesis to Berners-Lee's scepticism towards the Internet's achievements up to today. Berners-Lee's view of the online world has theoretically been portrayed as 'double-edged' (Malecki and Moriset, 2007), or a 'spiky' one (Florida, 2005). Cairncross' concept is based on the idea that new information and communication technology (ICT) is rapidly obliterating distance as a relevant factor in how we conduct business and our personal lives. Conversely, Warf (2013) posits that such views hinge on simplistic and utopian technological determinism that ignores the complexity associated with economic and social structures. 'Being there' (Gertler, 2003) and benefiting from the 'buzz' (Storper and Venables, 2004) still seems important for people, and technology cannot (yet) replace human interaction, although it can attempt to bridge distances virtually.

The idea of the Internet intuitively steers the mind towards a borderless and virtual world, with online users and online service providers vastly spread geographically and with a shortened psychic distance between the Global North and the Global South. Along with the permeation of the Internet in our society, online service industries have thrived in many parts of the world, including in some emerging markets. ICT and online services have been said to be important factors for 'global shifts' (Dicken, 2011). Some authors (Hruby and Coulter, 2015) argue that the Internet is transforming low income contexts like Africa into global powerhouses thanks to online services like m-pesa and crowdsourcing apps. McKinsey (2013) suggests a similar outlook for Internet's transformative potential in Africa and argue that 'there is a growing wave of innovation as entrepreneurs and large corporations alike launch new web-based ventures'. More sceptically, Warf (2001a) claims that 'the Internet creates and reflects a distinct spatial structure interlaced with, and often reinforcing, existing relations of wealth and power.' Warf implies that rich countries will continue to dominate also in the online world. Apparently, there are diverging opinions between scholars whether the Internet and its online services will equalise the economic geography between the Global North and the Global South.

In the chapter we analyse the spatial inequality of online services from firstly the demand side, where we study the usage of online services, measured as 'Internet penetration' with data from The World Bank. As a complement to Internet penetration we provide in-depth qualitative analysis of the user geographies of Twitter, Facebook and Wikipedia in the chapter. Secondly, from the supply side, we look at geographical origin and internationalisation of the world's largest online service providers on the firm level. The demand and supply of online services are important components from an economic development perspective. From previous studies (Hilbert, 2010; Brynjolfsson and McAfee 2014) we know that ICT has a strong impact on economic growth. Unless we have a clear view of spatial disparities in terms of online service demand and supply it is difficult for policy makers, particularly in lagging nations, to take appropriate action to enable more inclusiveness in the information society that can bring sustainable economic growth.

The two research questions of the chapter are: 1) where (geographically) are online services consumed and produced today?; 2) does the current geographical pattern of online demand and online supply reinforce already existing geographical disparities of economic power between nations, or does it have equalising effect, meaning that countries in the Global South are catching up?

The relevance of online services and online service providers

The most obvious online services are online social media services provided by firms such as Facebook and Twitter. Apart from social media, there are other online services as games, music streaming, news and information related services for example search engines. If we contemplate the nature of the service itself we can conclude that it can be more or less digitised and dependent on hardware. The extreme case is online services that *only* exist online, for example online search engines. But there is also a grey zone of online services which could also be consumed both in the online and in the physical space, for example bank services and broker services for tangible goods, like eBay. For the purpose of the chapter we focus on digitised and intangible services, which are *mainly* consumed online.

One adequate and widely accepted measure of the demand side of online services is 'Internet penetration'. Internet penetration is an indicator of how many people in a country have access to the Internet, and it is measured as percentage of the total population. It is an international recognised measure, collected in almost all countries, and published by ITU (2014) and The World Bank (2015).

On the supply side of online services we know that most online services are produced by firms, often referred to as 'online service providers'. Nevertheless, online services are also provided by non-governmental organisations, and by state-owned entities, for example e-government projects. According to OECD's Internet Economy Outlook there are ten online service providers among the world's 300 largest ICT firms, and this group of firms is becoming increasingly important in the world economy (OECD, 2012). Google is the world's 52nd largest firm (Forbes, 2015) and the relevance and influence of online service providers is rapidly increasing. In national commerce registries online service providers are often classified as software- or media firms. This makes it difficult to filter out which ones are online service providers. Fortunately, there are specialised analytical firms, like Statista, which have collected data of the world's largest online service. In the chapter we use this data and we also include examples from different geographical regions in our discussion via research of secondary data.

Due to the scarcity of coherent data, a limitation of the chapter is that it does not give a complete picture of online service providers from all geographical regions. However, it covers the world's largest online service providers, which we argue is an important indicator when assessing geographical disparities of the production of online services.

The geography of the Internet – the demand side

Internet penetration and the digital divide

There is an abundance of research and reports measuring Internet with supply-side data, for instance the number of domain names and Internet hosts, bandwidth, and so on (Zook, 2001). These variables are relevant, but different from our focus is on Internet penetration. There are approximately 3 billion Internet users worldwide (ITU, 2014) out of a world population of approximately 7.5 billion, hence an Internet penetration of 40 per cent globally. Between 75–80 per cent of households have access in the developed countries, around 30 per cent in developing countries, and 5 per cent in less-developed countries. There is still a substantial gap between developed and developing countries and more than 90 per cent of the people who are not yet using the Internet are from the developing world (ibid.). In Africa, around 19 per cent of the population was online in 2014, up from 10 per cent in 2010, and from 1.68 per cent in 2003. It is noteworthy that a much larger proportion of the population is online in the Global North than in the Global South, albeit there are nominally more people online in the Global South. China is the country with the largest number of Internet users (641 million), followed by the US (270 million) (ibid.).

Africa has the lowest Internet penetration rate in the world, although the average annual growth of users has been particularly high in some countries (Warf, 2001b). In Africa, only 10.7 per cent of the households are connected to the

Internet (see Table 4.1), and almost half of the Internet-connected share of the population goes online through shared connections, for example through Internet cafés. Household Internet access in Africa continues to grow at double-digit rates, albeit unevenly. Morocco has the highest Internet penetration in Africa (56 per cent) whereas Burundi has the lowest (1.3 per cent) (World Bank, 2015).

The 'digital divide', a term for the inequality of ICT access (Castells, 2002; Van Dijk 2006; Hilbert, 2010), is in Table 4.1 expressed as Internet penetration with sub-Saharan Africa as a reference point. It is interesting to note that the gap (here measured in per cent units) between sub-Saharan Africa and the rest of the regions is actually larger today than 10 years ago. For example, between Africa and Asia and Pacific the gap has widened with 17.5 per cent units during the last decade. Thus far, we conclude that the spread of the Internet has not been equal between geographical regions and that the digital divide seems to persist.

The geographies of Twitter and Wikipedia

Twitter is a social media message service which has been active since 2006, and processes approximately 500 million tweets per day (Twitter, 2015). Hawelka *et al.* (2014) have used geo-located Twitter as a proxy for measuring global mobility, and found that the Twitter network forms spatially cohesive regions reflecting the regional division of the world. In similar vein, Graham (2013a) argues that the Internet is a reflection of the real networks of people's physical worlds so that one could question the transcending effects of the Internet. In Graham's *et al.* (2013) Twitter map, the flagrant inequality in the geography of Twitter content is accentuated. However, they argue that in contrast to many online platforms with producer dominance from the Global North, Twitter displays significantly different geographies. For example, the six largest countries in terms of user tweet production through Twitter are: (1) US, (2) Brazil, (3) Indonesia, (4) UK, (5) Mexico, and (6) Malaysia. Interestingly, this is quite a heterogeneous group of

Table 4.1 Internet penetration in different regions. Authors' elaboration with data from ITU

	2005	*2015**	*Increase in percent units 2005–2015*	*Gap with Africa 2005*	*Gap with Africa 2015*	*Change 2015/2005 in percent units**
Africa	1.0	10.7	9.7	N/A	N/A	N/A
Arab States	9.5	40.3	30.8	8.5	29.5	21.0
Asia and Pacific	11.8	39.0	27.2	10.8	28.2	17.5
CIS	11.1	60.1	49.0	10.0	49.3	39.3
Europe	42.0	82.1	40.1	40.9	71.3	30.4
The Americas	32.7	60.0	27.4	31.6	49.3	17.6

Note: * Estimate. Rounded values. N/A: Not applicable. Regions in this table are based on the ITU BDT Regions, see: www.itu.int/en/ITU-D/Statistics/Pages/definitions/regions.aspx

Source: ITU World Telecommunication/ICT Indicators database.

countries. This indicates that countries from the Global South are indeed contributing to the online information flow. Thirty per cent of all tweets are from the US, and 23 per cent of the total number of accounts is within the US. The findings by Leetaru *et al.* (2014, n.p.) take a similar view. They conclude that 'there appears to be only weak geographic affinity in communicative link formation in that users retweet and reference users far away nearly as often as they do those physically proximate to them.' They estimate that approximately half of the tweeted news-related messages are about events close to people, and thus half from greater distance. They also argue that emerging markets such as the Middle East and Eastern Europe account for some of Twitter's largest new growth areas, while Indonesia, Western Europe, Africa, and Central America have high proportions of the world's most influential Twitter users. The Economist (2015a) notices that Saudi Arabia is one of the most tweeting countries relative to its Internet penetration. Saudi's most followed account (11.4 million followers) is that of a Saudi cleric. These findings are somewhat contradictory. On the one hand, Twitter usage is less restricted to geographical proximity, and quite geographically diverse on the demand side. On the other hand, the Saudi example shows restrictiveness in terms of psychic distance, for example, religious, language and cultural proximity. We conclude that Twitter does not necessarily create new virtual networks of different people although it may connect people from the same social or ethnical origins over wider geographical spans. This means that there is a bridging effect, but not necessarily between people with various backgrounds.

The geography of Wikipedia has been studied by Graham (2013b) and reveals a divided pattern between the Global North and the Global South. The US, Germany, UK and France are dominating the number of edits with over a million edits each quarter. Africa is far behind: only Egypt, Algeria, Morocco, Tunisia and South Africa make notable contributions to Wikipedia content. There are more edits emerging from Hong Kong than there are from the whole African continent. Also, in terms of content, the online encyclopaedia remains uneven and clustered, with a lot of information about only a few places (Graham *et al.*, 2013). Yet again, there is a considerable online information gap between different geographical places.

The geography of Facebook

Facebook is currently (July 2015) the world's largest web community with 1.49 billion users (Facebook, 2015a). In terms of number of users, the closest competitor is the Chinese Tencent, with approximately 830 million accounts, followed by WhatsApp's 700 million (The Economist, 2015b). It is interesting to note that China's Internet penetration rate has increased from 7 per cent to more than 45 per cent in one decade. Both Facebook and Twitter are censored in China, and it could be discussed whether this censorship has actually been an important trigger for online use and online entrepreneurship in China. The government involvement in China is, however, also challenging for the development of the industry due to the difficulty of combining policies of techno-nationalism, information control and social movements. These are difficult to control simultaneously (Ernkvist and Ström, 2008).

Unsurprisingly, Facebook users are not evenly spread around the world, and are particularly underrepresented in Africa. According to Facebook, their penetration is around 10 per cent in Africa in comparison to 19 per cent worldwide (Facebook, 2015b). There is a strong positive correlation between economic strength (GDP per capita) and Internet and Facebook penetration in Africa, but only a weak positive correlation between Facebook penetration and Internet penetration (Wentrup *et al.*, 2015). It is interesting to note that Facebook penetration is actually higher than Internet penetration in some countries, for example Botswana. This phenomenon indicates that Facebook is consumed in spaces which provide shared Internet access, like Internet cafés and workplaces. Figure 4.1 shows Facebook penetration as dark blue circles, Internet penetration as light

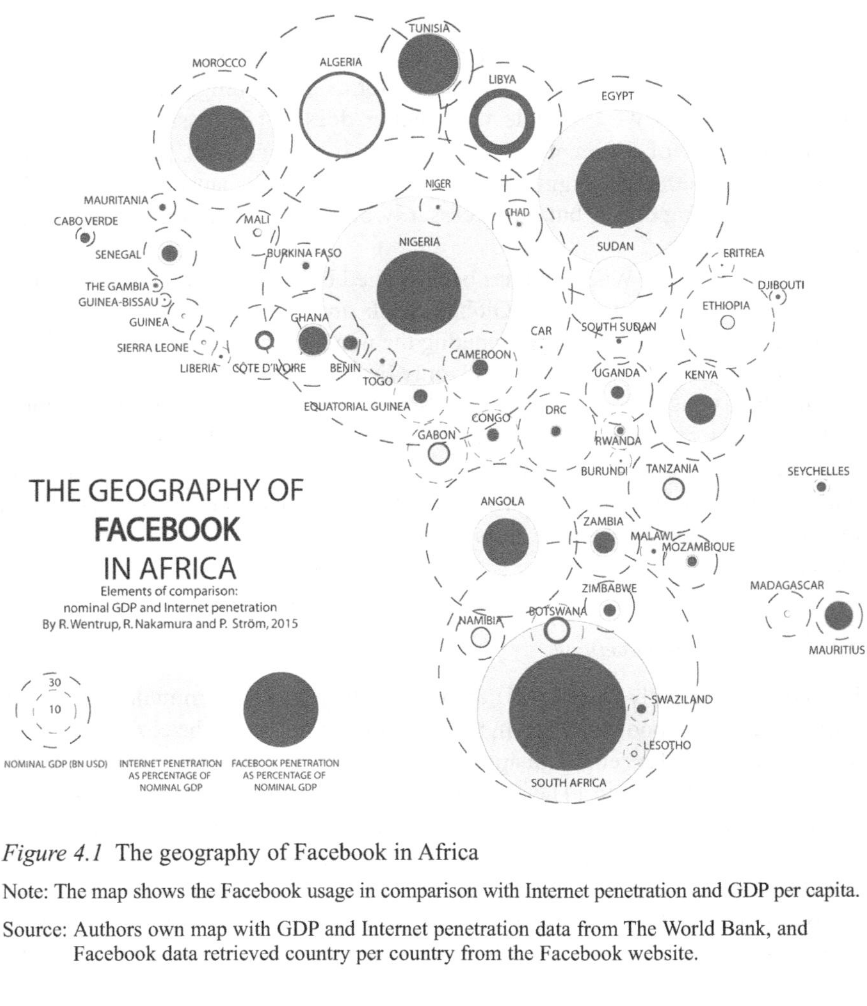

Figure 4.1 The geography of Facebook in Africa

Note: The map shows the Facebook usage in comparison with Internet penetration and GDP per capita.

Source: Authors own map with GDP and Internet penetration data from The World Bank, and Facebook data retrieved country per country from the Facebook website.

blue circles and nominal GDP as grey dashed circles. The Internet and Facebook landscape is heterogeneous. The North African countries are performing better than their sub-Saharan on all measures shown in the map. It is interesting to note that a few relatively wealthy African countries are not filling out their GDP circles with neither Internet nor Facebook penetration particularly well, like Angola. Although these countries are fairly rich, they have not managed to succeed in distributing online access for their citizens. In contrast to these countries, we notice that Kenya, despite being fairly poor, have impressive Internet and Facebook penetration. It is also interesting to note that the sub-Saharan countries, despite being so close geographically, differ so substantially digitally. This shows the online heterogeneity within regions and could question agglomeration effects within regional digital development. Affordability (cost of Internet access in comparison to GDP per capita), often as a consequence of lack of competition in the telecom market, and literacy are said to be two of the key reasons behind the heterogeneity (Oyelaran-Oyeyinka and Lal, 2005).

The geography of online service providers – the supply side

Dominance by US (Silicon Valley), China, and global cities

As mentioned in the beginning of the chapter, most online services are produced by firms called online service providers. Let us now look at the geographical origins of the world's largest online service providers to find out whether there are any particular patterns. At a first glance (left columns in Table 4.2) we find that there is dominance by firms from the US and China. Nearly nine out of ten of the largest online service providers are either from the US or China. The table also reveals the dominance by US firms over China. US firms account for almost six out of ten positions of the world's largest online service providers.

It is interesting to note that there are no European, African, or Latin American online service providers among the largest ones, but that both South Korea and Japan each have one firm on the top list. If we compare this list with the largest firms in the world (right columns in Table 4.2), we can see that the dominance by the US is stronger in the online service ranking than it is in the ranking of the world's largest firms. In the latter list there is more geographical diversity. On an aggregate level the table clearly demonstrates the geographical concentration of the world's largest online services to the US, but closely followed by China. The US firms have also a far better global reach in terms of markets, relative to the Chinese firms where the main base lies within the domestic market. However, with increased internationalisation among Chinese firms this might change in the future.

In the Global North, Silicon Valley stands out as the main hub for the Internet economy and most of the dominating US-based firms come from this area (Zook, 2002; Warf, 2013; Piscione, 2013). The concept of 'being there', narrow social relations and tacit knowledge, even in a virtual industry such as the Internet industry, is apparently relevant.

Table 4.2 The world's largest online service providers and the world's largest firms from various sources

World's largest online service providers				*World's largest firms*			
*Source: Statista**	*Origin*	*Source: Wikipekia***	*Origin*	*Source: Forbes****	*Origin*	*Source: Wikipekia*****	*Origin*
Google	US	Amazon	US	ICBC	China	Wal-Mart	US
Alibaba	China	Google	US	China Construction Bank	China	Sinopec	China
Facebook China	US	JD.com	China	Agriculture: Bank of China	China	China National Petroleum Corporation	
Amazon.com	US	eBay	US	Bank of China	China	Royal Dutch Shell	Netherlands
Tencent	China	Tencent	China	Berkshire Hathaway	US	ExxonMobil	US
eBay	US	Facebook	US	JPMorgan Chase	China	Saudi Aramco	Saudi Arabia
Baidu	China	Alibaba	China	ExxonMobil	US	BP	UK
Priceline.com	US	Priceline.com	US	PetroChina	China	State Grid Corporation of China	China
Salesforce.com	US	Expedia.inc	US	General Electric	China	Samsung Group	South Korea
JD.com	China	Rakuten	Japan	Wells Fargo	US	Vitol	Netherlands and Switzerland
Yahoo!	US	Netflix	US	Toyota Motor	Japan	Toyota	Japan
Netflix	US	Baidu	China	Apple	US	Glencore Xstrata	Switzerland
LinkedIn	US	Yahoo!	US	Royal Dutch Shell	Netherlands	Volkswagen	Germany
Twitter	US	Salesforce.com	US	Volkswagen	Germany	Total	France
Yahoo! (Japan)	US	Groupon	US	HSBC	UK	Chevron	US
Rakuten	Japan	LinkedIn	US	Chevron	US	Berkshire and Hathaway	US
NetEase	China	NetEase	China	Wal-Mart	US	Apple	US
Navar	South Korea	Twitter	China	Samsung Electronics	South Korea	China Railway Corporation	China
Vipshop	China	ASOS.com	UK	Citigroup	US	Phillips 66	US
		Tripadvisor	US	China Mobile	China	E-ON	Germany
		Yandex	Russia	Allianz	Germany	Gazprom	Russia

Notes: * By market value of the largest Internet companies worldwide as of May 2015. Only ranks 19 top firms. Source: www.statista.com/statistics/277483/market-value-of-the-largest-internet-companies-worldwide/

** By revenue and market capitalization. Source: https://en.wikipedia.org/wiki/List_of_largest_Internet_companies. Only ranks 21 top firms.

*** By metric of sales, profits, assets and market value. Source: www.forbes.com/global2000/

**** By consolidated revenue. The list is limited to companies with annual revenues exceeding 100 billion USD. Source: https://en.wikipedia.org/wiki/List_of_largest_companies_by_revenue

The Global North has generated multiple Internet and start-up clusters in the last decade, and many countries have often tried to replicate the Silicon Valley formula. These clusters are often filled with incubators which provide services such as physical infrastructure, administrative support, technical support and access to financing, legal advice and networking to facilitate the development of new businesses (Rosa, 2014). Most are located in the largest metropolitan areas in capital cities, for example London, Paris, Stockholm (The Economist, 2013), and Tel Aviv (The Economist, 2014). This pattern shows similarities with other advanced service industries that are to be found in primary urban agglomerations (Gertler, 2003; Ström, 2006).

The online service industry has so far reinforced, not diminished, agglomeration effects and the role of global cities (Warf, 2013). Not only are the most important online service providers located in the global cities, but they are most often also in central, prestigious districts within the cities (Wentrup and Ström, 2013). Zook (2002) suggests that venture capital has played an important role in the concentration of the Internet industry in a few places. The relevance of geographical proximity and the local buzz (Bathelt *et al.*, 2004) should not be underestimated. Online services seem to enable and facilitate the creation of relational proximity, but online service providers are sensitive to the advantages brought about by geographical proximity (or spatial agglomeration effects) as many other types of producers of high-tech and advanced informational services (Amin and Cohendet, 2005; Balland *et al.*, 2011)

Apart from the US, few countries in the Global North have been able to give birth to online service providers serving a large user base spread over a vast geographical area. In Europe the UK has produced most online service providers (Ahmed, 2015). But also in smaller countries like Sweden online service providers like Skype (acquired by Microsoft) and Spotify have emerged. Israel is another example with the online traffic service Waze. Waze was sold to Google in 2013. Both Skype and Waze show how large US online service providers expeditiously have managed to acquire small fast growing online service providers. Since 2001, Google has acquired more than 180 firms, and Facebook more than 50, of which WhatsApp is the largest (The Economist, 2014). Most of the acquired firms are American, and a few are from other countries in the Global North. Only one firm from the Global South (India) has so far been acquired by either Facebook or Google. The aggressive acquisition strategy could partly explain the US dominance in the online industry. To our knowledge Chinese online service providers have not yet been so aggressive in acquiring fast growing online service providers (Yu, 2015), although Alibaba made a few acquisitions in 2014 (Bischoff, 2014). For online service providers in the Global South it could therefore be considered to put more emphasis on acquiring or collaborating with online service providers. Through such acquisitions, mergers or partnerships they could capitalise on the online internationalisation experience of the Global North online service providers and thereby gain access to established networks in the these markets. Online internationalisation is a fairly new research field and has been observed by a limited number of researchers (see Kim, 2003; Yamin and Sinkovics, 2006; Forsgren and

Hagström, 2007; Gabrielsson and Gabrielsson, 2011; Sinkovics *et al.*, 2013). When studying online internationalisation it is clear that the success of the firms is linked to the entrepreneurs and their networks (Jones and Coviello, 2005; Vahlne and Johanson, 2013) and geographical location (Rusten and Bryson, 2010). For example, many of the US online services providers are located in the limited geographical area around Mountain View, California, and most of the successful Swedish online service firms have emerged from the city centre of Stockholm via a small and closely connected group of entrepreneurs. In order to grow and expand rapidly, entrepreneurs need access to savvy programmers and knowledgeable highly skilled staff; innovation capabilities; and a mix of investors, academics and entrepreneurs with international networks (Piscione, 2013). These assets are critical for international success and they are today, to a large extent, concentrated in restricted geographical spaces within a few global cities in the Global North.

Promising exceptions from the Global South, but not sufficiently international

So are people around the world only using online services developed in the Global North? Not quite. As Table 4.2 shows, many of the world's largest online service providers are from China. Nonetheless, there is a significant difference if we compare the target geographies of online service providers from the US and China. The online services with a geographically spread target audience are almost all from the US. Yet, people in the Global South do use domestically developed online services adapted to the local needs, particularly for purchasing goods and looking at news. In India, for example, there are successful local online service providers, such as Snapdeal.com and Ola Cabs (The Wall Street Journal, 2014). The former is an online marketplace with more than 25 million users and the latter is an online taxi-hailing service. However, these online service firms only serve Indian consumers and have as yet not been internationalised. Nine out of the ten top visited online services in India are all provided by US firms (see www.alexa.com/topsites/countries/IN) and the tenth largest is an Indian news website with limited interest for an international audience.

The pattern is a bit different for the Chinese online service firms. An important difference with India though is that the Chinese online service providers have succeeded in scaling up and attracting a large part of the Chinese online crowd, whereas the Indian online service providers are still relatively small, which might be due to different market structures and socio economic development. All of the top ten most visited online services in China are provided by Chinese online service providers (see www.alexa.com/topsites/countries/CN). Contributing factors for this pattern are a sudden rise in consumer demand as a result of China's economic growth; success in attracting FDIs; and first mover advantage (Anwar, 2013) on the large home market are some important drivers. Chinese firms have thrived alongside a vast fibre network roll-out in major Chinese urban areas. Because of urbanisation and demand for new services (Chen *et al.*, 2013) the Chinese online service providers have bloomed in a prosperous climate. China's hybrid economy with flagrant market imprecations-coupled censorship, has protected China from some of the most aggressive American competitors, and thus opened up growth potential for domestic firms. The challenge for future development will depend on whether Chinese firms can successfully internationalise into other markets, or if they are just a result of shielded domestic markets. If the latter is the case it could also help to sustain differences between mature and controlled emerging markets (Alvstam *et al.*, 2015).

Whereas the Global North firms have followed a regional geographical pattern (Kim, 2003) and expanded internationally early after inception, the Chinese firms have internationalised via the Chinese diaspora, and more restrictively relative to the geography (Anwar, 2013). Unfortunately, the Indian online service providers do not seem to come far with their internationalisation. Although Bangalore in India has become a centre of code programming, and has succeeded in producing traditional software – and consultancy firms (for example, Little Eye Labs, which was sold to Facebook, and Infosys) – the city has not yet produced international online service ventures reaching an international audience.

From the Global South there is one often cited African case in the money transfer and micro-financing online service from Kenya, called m-pesa (The Economist, 2013c; The Guardian, 2014), which is owned by Vodafone (UK). M-pesa has expanded geographically to five or six countries in Africa as well as to Afghanistan and a few countries in Eastern Europe. More than 15 million people are using the service, and Nairobi has, together with Kigali, some of Africa's most dynamic start-up communities. Hitherto and unsurprisingly given the aim of the service, m-pesa has not attracted the users in the Global North. Again, this is an example of how online services produced in the Global South do not target and reach people in the Global North, and remain internationally restricted.

Why do online services from the Global South not reach the Global North?

There is an absence of entangling effects on the supply side in the south-north direction. Even though countries in the Global South are rapidly increasing their Internet connectivity rates, they are still irrelevant in a global context in the Internet economy (Sonaike, 2004). Global South countries are already in a techno-dependency, where they are constantly lagging a few steps behind in both consumption and production within the online service industry. We suggest three propositions aiming to explain this lag in comparison to the Global North:

Path dependency

It is suggested that path dependency and legacy in trade patterns have effect in the online service economy. Global South countries are stuck with unfavourable terms of trade in their extant trade patterns with the Global North, and have little export experience of knowledge-intensive services. International flows of online services do not automatically alter this pattern.

Lack of ecosystems and incubators

The second proposition we put forward is the absence of efficient ecosystems with access to entrepreneurs, capital and international networks, including connections with the global cities. The brain drain of Internet talent to the Global North and insufficient number of diaspora returnees are contributing factors.

Domestic lock-in

We argue that the large domestic populations in the Global South, for example in China and India actually hinder an early international expansion because the home market is sufficiently large. The effect is a domestic lock-in. Hence, the firms focus too much on local adaptions to home markets and end up in a domestic lock-in, which lowers their potential for international expansion. Additionally, the application areas of the online services produced in the Global South are sometimes less relevant for the Global North. For example micro-financing payment systems, developed for the African needs are probably not scalable in the Global North

An online service from the Global South could potentially go viral in the Global North's online space in the near future. In the gaming industry we have seen the good example of Vietnamese Flappy's global success. The readiness of technology, capital and know-how exists in some districts, and the Global South has already proved that it has been able to produce online services reaching a large audience, albeit geographically limited.

Concluding remarks

Although we are today witnessing fascinating global shifts, where countries in the Global South are starting to invest and trade with each other, and where countries in the Global South are making investments in the Global North (Hermelin *et al.*, 2011), the geography of online services is still a mirror of the past. Most of the demand, and most of the supply, comes from the high-end urban districts in the Global North, which is a reflection of the global economy and its underpinning power structures. We conclude that there is still an important digital divide on both the demand and supply side, although there are promising examples from the Global South, led by China.

This analysis may seem slightly pessimistic, but there are attention-grabbing signs that the online service industry may be one of the first which will witness a more even playing field between the Global South and the Global North. One strong indicator is the speed at which the online service geography is changing. It has quickly gone from being a completely Global North affair to include the Global South. In comparison, this global transition has taken a much longer time in other industries. Internet penetration rates are increasing, with double-digit growth figures in some countries the Global South, and e-commerce is burgeoning all over the planet. The examples from the Global South mentioned in the chapter are important indicators that online service providers can, at least technically, scale and spread into the Global North. We argue that such a development is important, not only for the growth of the online service industry in the Global South, but also to make the online service geography more even, diverse and dynamic.

A first natural step for online service providers in the Global South is to serve the online needs in their home market before entering the global market. This phase is already *en route*. Yet, it is crucial for Global South firms not to get stuck in the home market phase too long. They need to grow internationally early after inception, and become competitive on the global market, and thus counterbalance the dominance of the Global North, to avoid being left in a permanent state of domestic lock-in, and techno-dependency with lagging economic development as a negative side effect. To avoid such a scenario, developing countries need to ensure that they keep talent, and concentrate entrepreneurs, capital, investors and academics into a few dynamic urban areas and grow from there, to avoid scattering their assets geographically. They may also seek to acquire or collaborate with online service providers in the Global North and thus get access to important networks. The combination of regional policy development and firm strategy is crucial in knowledge-driven economies (Ström and Wahlqvist, 2010). They need to adapt a rigorous internationalising mind-set. Furthermore, entrepreneurs in the Global South must strive to learn and integrate online internationalisation at the outset of their entrepreneurial endeavours, in order to compete with international online service providers. The priority for regional policymaking must be to drive demand and thus ensure that (at least) a critical mass of their citizens are online, and in parallel lay the foundation for an entrepreneurially dynamic ground in order to transform some of the citizens into online service producers ready to compete internationally.

Further research

We suggest three main ideas for further research. Firstly, as mentioned in the introduction of the chapter, research of online services is cumbersome due to the lack of aggregated data. Therefore there is need for the international research community to jointly collect reliable data about online services (both on the demand and supply side). As shown in the chapter, the available data is still scarce and scattered, which makes researchers (us included) turn to crowd-sourced data such as Wikipedia. The online service providers themselves, as well as telecom operators have an important role to play herein in sharing their data with the research community and vice versa.

Secondly, the total online population in the Global South is in nominal terms substantially larger than the one in the Global North, although the proportion is higher in the Global North. It would be interesting to disentangle the causes behind the low number of online service providers from the Global South, and to understand at what point an online crowd in a geographical space becomes entrepreneurial, and transform itself from a consumer into producer in the online service industry.

Thirdly, it would be interesting to see how online service providers in the Global South can tap into the possibilities Internet of Things (IoT). The potential of interconnectedness through IoT will help to transform the market for various online services. Firms within the Global South could potentially use this development to make use of global value chains through connecting low cost production of goods and services. Given their large domestic markets in combination with rapid economic growth, online development can be used as a platform for connecting the society to the global potential of sharing data and building completely new business models for the Global South.

References

Ahmed, M. (2015). Billion-dollar tech start-ups on the increase in Europe. Available at: www.ft.com/intl/cms/s/0/bf7b3dd4-127d-11e5-8cd7-00144feabdc0.html#axzz3h0uLRogV, [accessed 20 July 2015].

Alvstam, Claes G., Ström, P. and Wentrup, R. (2016). Heterogeneous economic space in a global archipelago – An economic geography perspective of emerging markets. In: Merchant, H. *et al.* (eds) *Handbook of Research on Emerging Markets*, Cheltenham: Edward Elgar.

Amin, A. and Cohendet, P. (2005). Geographies of Knowledge Formation in Firms, *Industry and Innovation*, 12(4), pp. 465–86.

Anwar, S.T. (2013). Icarus paradox or verto in prosperum: Chinese Internet firms, business models and internationalization. *The Journal of World Investment & Trade*, 14(3), pp. 532–55.

Balland, P-A., de Vaan, M. and Boschma, R. (2013). The dynamics of interfirm networks along the industry life cycle: The case of the global video game industry, 1987–2007, *Journal of Economic Geography*, 13(5), pp. 741–65.

Bathelt, H., Malmberg, A. and Maskell, P. (2004). Clusters and knowledge: local buzz, global pipelines and the process of knowledge creation, *Progress in Human Geography*, 28(1), pp. 31–56.

Bischoff, E. (2014). *Jack Ma's Shopping Spree: Here's everything Alibaba invested in or acquired in 2014*, [online] Available at: www.techinasia.com/alibaba-investments-2014 [accessed 10 July 2015].

Brynjolfsson, E. and McAfee, A. (2014). *The Second Machine Age: Work Progress and Prosperity in a Time of Brilliant Technologies*. New York: WW Norton Company Inc.

Cairncross, F. (2001). *The Death of Distance 2.0: How the communications revolution will change our lives*. London: Texere.

Castells, M. (2002). *The Internet Galaxy*. Oxford: Oxford University Press.

Chen, M., Liu, W. and Tao, X. (2013). Evolution and assessment on China's urbanization 1960–2010: under-urbanization or over-urbanization? *Habitat International*, 38, pp. 25–33.

Dicken, P. (2011). *Global Shift*, 6th ed. London: Sage Publications.

The Economist (2013). *A Tale of Two Ecosystems*. Available at: www.economist.com/blogs/schumpeter/2013/10/start-ups-europe [accessed 12 February 2015].

The Economist (2014). *The Scale Up Nation*. Available at: www.economist.com/news/business/21636111-israel-trying-turn-its-davids-goliaths-scale-up-nation [accessed 29 March 2015].

The Economist (2015a). *Why Saudis are Ardent Social Media Fans*. Available at:

www.economist.com/blogs/economist-explains/2015/03/economist-explains-21?fsrc=scn/fb/wl/ee/tr/saudisareardentsocialmediafans [accessed 10 April 2015].

The Economist (2015b). *What's Up*. Available at: www.economist.com/blogs/graphicdetail/2015/03/messaging-apps [accessed 30 March 2015].

Ernkvist, M. and Ström, P. (2008). Enmeshed in games with the government: governmental policies and the development of the Chinese online games industry. *Games and Culture*, 3(1), pp. 98–126.

Facebook (2015a). Facebook *newsroom*. Available at http://newsroom.fb.com/company-info/ [accessed 30 July 2015].

Facebook (2015b). *Connecting 100 Million People in Africa*. Available at www.facebook.com/business/news/connecting-100-million-people-in-africa [accessed 15 March 2015].

Florida, R. (2005). The world is spiky, *The Atlantic Monthly*, 51 (October), pp. 48–51.

Forbes (2015). *Global Leading 2000 Companies*. [Online] Available at: www.forbes.com/global2000/ [accessed 13 April 2013].

Forsgren, M. and Hagström, P. (2007). Ignorant and impatient internationalization? The Uppsala model and internationalization patterns for Internet-related firms. *Critical Perspectives on International Business*, 3(3), pp. 291–305.

Friedman, T.L. (2005). *The World is Flat: A brief history of the twenty-first century*. New York: Farrar, Straus and Giroux.

Gabrielsson, M. and Gabrielsson P. (2011). Internet-based sales channel strategies of born global firms. *International Business Review*, 20(1), pp. 88–99.

Gertler, M. (2003). Tacit knowledge and the economic geography of context or the undefinable tacitness of being (there). *Journal Economic Geography*, 3(1), pp. 75–99.

Graham, M. (2013a). Geography/internet: ethereal alternate dimensions of cyberspace or grounded augmented realities? *Geographical Journal*, 179(2), pp. 177–82.

Graham, M. (2013b). *Geographies of Information in Africa: Wikipedia and User-Generated Content,* R-Link: Rwanda's Official ICT Magazine, Kigali: Rwanda ICT Chamber, pp. 40–41.

Graham, M., Stephens, M. and Hale, S. (2013). Mapping the geoweb: a geography of Twitter. *Environment and Planning A*, 45(1), pp. 100–102.

The Guardian (2014). *Watch out M-Pesa, Equity Bank wants to transform mobile money in Kenya*. Available at: www.theguardian.com/global-development-professionals-network/2014/oct/02/mpesa-safaricom-equity-bank-financial-inclusion-kenya [accessed 30 March 2015].

Hawelka, B., Sitko, I., Beinat, E., Sobolevsky, S., Kazakopoulos, P. and Ratti, C. (2014). Geo-located Twitter as proxy for global mobility patterns. *Cartography and Geographic Information, Science*, 41(3), pp. 260–271.

Hermelin, B., Demir, R. and Verhagen, H. (2011). Reconsidering the geographical flows of the global knowledge economy. The example of Indian ICT-based knowledge flows to Sweden. *International Journal of Digital Society*, 2(2).

Hilbert, M. (2010). When is cheap, cheap enough to bridge the digital divide? Modeling income-related structural challenges of technology diffusion in Latin America. *World Development*, 38(5), pp. 756–70.

Hruby, A. and Coulter, R. (2015). *Four trends accelerating entrepreneurship in Africa*. Available at: https://agenda.weforum.org/2015/07/four-trends-accelerating-entrepreneurship-in-africa [accessed 8 July, 2015].

ITU. (2014) (International Telecommunication Union)., 2013. *Measuring the Information Society 2014*, ITU, Geneva, Switzerland. [Online] Available at: www.itu.int/pub/D-IND-ICTOI-2014 [accessed 14 July 2014].

James, J. (2008). Digital preparedness versus the digital divide: a confusion of means and ends. *Journal of the American Society for Information Science and Technology*, 59(5), pp. 785–91.

Jones, M.V. and Coviello, N.E. (2005). Internationalisation: conceptualising an entrepreneurial process of behaviour in time. *Journal of International Business Studies*, 36(3), pp. 284–303.

Kim, D. (2003). The internationalization of US internet portals: does it fit the process model of internationalization. *Marketing Intelligence & Planning*, 21(1), pp. 23–36.

Leetaru, K.H., Wang, S., Cao, G., Padmanabhan, A. and Shook, E. (2014). *Mapping the global* Twitter *heartbeat: the geography of* Twitter. Available at: http://firstmonday.org/article/view/4366/3654 [accessed 14 March 2015].

LeWeb (2014). *In conversation with Sir Tim Berners Lee.* Available at: www.youtube.com/watch?v=PowqvYH8tPU [accessed 15 February 2015].

Malecki, E.J. and Moriset, B. (2007). The paradox of a 'double-edged' geography: local ecosystems of the digital economy. In: Malecki, E.J. and Moriset, B. (eds) *The digital economy: business organization, production processes and regional developments.* New York: Routledge, pp. 174–98.

McKinsey (2013). *Lions Go Digital: The Internet's transformative potential in Africa.* Available at: www.mckinsey.com/insights/high_tech_telecoms_Internet/lions_go_digital_the_Internets_transformative_potential_in_africa [accessed 15 May 2014].

OECD (2012). *Internet Economy Outlook.* Paris: OECD Publishing.

Oyelaran-Oyeyinka, B. and Lal, K. (2005). Internet diffusion in sub-Saharan Africa: a cross-country analysis. *Telecommunications Policy*, 29(7), pp. 507–27.

Piscione, D.P. (2013). *Secrets of Silicon Valley: What everyone else can learn from the innovation capital of the world.* Basingstoke: Palgrave Macmillan.

Rosa, C. (2014). Benchmarking of business incubators. *Benchmarking: An International Journal*, 21(6), pp. 1062–69.

Rusten, G. and Bryson, J.R. (2010). Placing and spacing services: towards a balanced economic geography of firms clusters social networks contracts and the geographies of enterprise. *Tijdschrift voor economische en sociale geografie*, 101(3), pp. 248–61.

Sinkovics, N., Sinkovics, R. and Jean, R-J. (2013). The internet as an alternative path to internationalization?. *International Marketing Review*, 30(2), pp. 130–155.

Sonaike, S. (2004). The internet and the dilemma of Africa's development. *International Communication Gazette*, 66(1), pp. 41–61.

Storper, M. and Venables, A. (2004). Buzz: face-to-face contact and the urban economy. *Journal of Economic Geography*, 4(4), pp. 351–70.

Ström, P. (2006). Internationalisation of Japanese Professional Business Service Firms: Dynamics of Competitiveness Through Urban Localization in Southeast Asia. In: Harrington, J.W. and Daniels, P.W. (eds) *Knowledge-Based Services: Internationalisation and Regional Development*, Aldershot: Ashgate.

Ström, P. and Wahlqvist, E. (2010). Regional and Firm Competitiveness in the Service Based Economy – Combining Economic Geography and International Business Theory, *Tijdschrift voor economische en sociale geografie*, 101(3), pp. 287–304.

The Wall Street Journal (2014). *SoftBank invests more than $800 million in India internet firms*. [Online] Available at: www.wsj.com/articles/softbank-invests-over-800-million-in-india-firms-1414471447 [accessed 20 March 2015].

Twitter (2015). *About Twitter*. Available at: https://about.twitter.com/company [accessed 26 March 2015].

Vahlne, J-E. and Johanson, J. (2013). The Uppsala model on evolution of the multinational business enterprise – from internalization to coordination of networks. *International Marketing Review*, 30(3), pp. 189–210.

Van Dijk, J.A.G.M. (2006). Digital divide research, achievements and shortcomings. *Poetics*, 34(4–5), pp. 221–35.

Warf, B. (2001a). Sequeways into cyberspace: Multiple geographies of the digital divide, *Environment and Planning B: Planning and Design*, 28(1). pp. 3–19.

Warf, B. (2001b). Uneven geographies of the African internet: growth, change, and implications. *African Geographical Review*, 29(2), p. 41.

Warf, B. (2013). Global geographies of the internet. Springer, e-book: DOI 10.1007/978-94-007-1245-4

Wentrup, R. and Ström, P. (2013). *Going viral across borders*. Conference paper presented at Reverse and intensified globalisation – the case of ICT, Flows and networks of trade, FDI, skilled labour, students, knowledge and ideas from India to regions in Europe. Norrköping, Sweden, 7 May, 2013.

Wentrup, R., Ström, P. and Nakamura, H.R. (2016). Digital oases and digital deserts in Sub-Saharan Africa. *Journal of Science & Technology Policy Management*, 7(1), 77–100.

World Bank (2015). *Internet users (per 100 people)*. Available at: http://data.worldbank.org/indicator/IT.NET.USER.P2 [accessed 14 March 2015].

Yamin, M. and Sinkovics, R. (2006). Online internationalisation, psychic distance reduction and the virtuality trap, *International Business Review*, 15(4), pp. 339–60.

Yu, S. (2015). *Baidu seen to be lagging rivals in mergers and acquisitions activity*. Available at: www.scmp.com/business/companies/article/1710435/baidu-seen-be-lagging-rivals-mergers-and-acquisitions-activity [accessed 3 July, 2015]

Zook, M.A. (2001). Old hierarchies or new networks of centrality? The global geography of the internet content market. *American Behavioral Scientist*, 44(10), pp. 1679–96.

Zook, M.A. (2002). Grounded capital: venture financing and the geography of the internet industry 1994–2000. *Journal of Economic Geography*, 2(2), pp. 151–77.

Part II

The changing role of services in the Global North

5 Knowledge-based services and economic development

Contributions to growth and resilience in European regions

Andrew Johnston and Robert Huggins

Introduction

In an increasingly service-dominated economy, attention is being turned towards how firms within this sector contribute to regional economic development (den Hertog, 2000; Gallouj, 2002; Miles, 2005; Wood, 2006; Gallaher and Petrusa 2006). Within the services sector, knowledge-intensive business services (KIBS) have been identified as important actors occupying a vital position in the modern economy (Hansen and Winther, 2010); as such, KIBS are increasingly described as important for economic development in terms of utilising high levels of human capital and turning this into innovative outputs. Consequently, these firms are viewed as both crucial to the innovation efforts of other firms as well as being innovators in their own right (Freel, 2006; Corrocher and Cusmano, 2014). As a result, terms such as 'knowledge-based' or 'knowledge driven' are increasingly used to describe the modern economy, and the competitive advantage of regions is now based on exploiting the knowledge assets they possess (Gertler, 2003; Jessop, 2000; Bryson *et al.*, 2000). Thus, knowledge is typically regarded as the most important factor of production as it 'is crucial in helping to create innovation which in turn stimulates economic growth and development' (Howells, 2002: 871)

Alongside these developments, the modern economy is also driven by consumer expenditure, with consumption-based services also important components of the modern regional economy (Johnston, 2009; Johnston, 2011). Thus, regional economies are increasingly viewed as the location of a number of pursuits based around consuming goods and services. As such, regeneration activities have typically led to ex-industrial sites being redeveloped into sites for leisure, entertainment, cultural and creative based pursuits (Evans, 2009; McCarthy, 2002) such as cinemas, ice rinks, shopping centres, restaurants, bars, art galleries, museums and sporting arenas, in order to improve both the economic prospects and the character of these areas (Totzer and Gigler, 2005).

The severe economic downturn experienced by advanced economies in the wake of the global financial crisis in the years since 2008 has led to increased interest in the resilience of regional economies, or their ability to adapt to economic shocks (Martin, 2011; Hassink, 2010). This chapter contributes to this burgeoning literature through examining the role of the service sector in regional economic resilience. The chapter adds to the extant literature by examining the relationship between employment changes in both knowledge-based services, characterised as business services, and consumption-based services, characterised as personal services, in order to assess how they contributed to the overall economic performance of European regional economies during the period of economic stagnation that occurred between 2008 and 2012. Consequently, the exploratory analysis presented in this chapter examines the role of employment in knowledge-based and consumption-based service sectors and the relationship between changes in employment in these sectors and overall economic performance in terms of GDP growth. Using quantitative data on employment levels and changes in regional GDP at the NUTS2 level, the explicit focus of the analysis is to examine the relationship between regional resilience and employment in the service sectors of the economy, both knowledge-based and consumption-based, and whether employment growth in either is associated with greater levels of regional resilience. As such, the analysis offers an overview of economic performance in the years following the 2008 Financial Crisis and subsequent recession and highlights the potential role of service-sector employment in the subsequent recovery following the recession.

The chapter is structured as follows: Section 2 outlines our theoretical and conceptual framework by examining the role of knowledge-based, business focused, services and consumption-based, people focused, services in the regional economic development process and the concept of regional resilience. Section 3 outlines the data sources used and Section 4 presents the analysis. Finally, Section 5 provides discussion, conclusions and implications of the findings.

Services and regional economic development

Within developed economies the services sector is dominant in terms of both employment and contribution to GDP, a pattern that became increasingly evident as the twentieth century progressed and is now manifest across Europe and most of the rest of the world (Jones, 2013; Uppenberg and Strauss, 2010; Daniels, 1991; Marato-Sanchez and Cuadrado-Roura, 2011). However, while the term 'services' has come to describe all tertiary sectors of the economy outside of the primary and manufacturing sectors, it cannot be considered to be a single homogenous group of firms but a diverse set of sectors (Illeris, 2007; Massey, 1984). As such, the services sector can be seen to be comprised of four broad groups: distributive services, producer services, personal services and social services (Browning and Singlemann, 1975), of which the focus is generally on the first three private sector, value generating, groups. The focus of this chapter is on two of these groups, producer services and personal services broadly conceptualised as knowledge-based services and consumption-based services (Johnston, 2009; Johnston, 2011).

The nature of these two broad service sub-sectors results in different output consumption patterns; typically, outputs from knowledge-based sectors have a larger, more geographically spread market, for example ICT services originating in Silicon Valley are aimed at the entire world market (Saxenian, 1994). Thus, knowledge-based services are more focused on 'traded services,' generating significant revenues from outside the region in which they were produced (Kaldor, 1970; Dixon and Thirlwall, 1975; Porter, 2003). Conversely, consumption-based sectors are increasingly based on experience goods, relying mostly on local demand for revenues (Lorentzen, 2009).

Employment in knowledge-based services

In advanced economies, mainly due to the increasing mobility of capital and globalisation, the nature of competitive advantage for regions has shifted away from being based on endowments of natural resources towards focussing on the knowledge resources of a location (Huggins and Izushi, 2007). This focus on knowledge resources has been accompanied by a corresponding focus on more innovative service sectors in the economy, i.e. knowledge-intensive business services. KIBS firms are providers of business to business services such as computer consultancy, R&D consultancy, accountancy, and legal practitioners (Miles, 2005). Their growing importance to regional economies is based on their role in terms of both providing and transferring knowledge and skills to their clients (Doloreux and Muller, 2007; Chadwick *et al.*, 2008; Doloreux and Shearmur, 2010; Huggins, 2011). Consequently, these firms are seen as 'bridges of innovation' (Corrocher and Cusmano, 2014: 4), acting as consultants and facilitators of business to business interaction through providing services that are typically tailored to the needs of their customer (Strambach, 2008). Thus, they are seen as both enablers of innovation among other firms and innovative firms in their own right (Miles, 2005; Leiponen, 2005).

Typically, therefore, KIBS firms are considered as integral components of regional knowledge networks and regional innovation systems, acting as bridging agents connecting knowledge sources and innovators (Den Hertog, 2000; Muller and Zenker, 2001; Miles, 2005; Koch and Stahlecker, 2006). KIBS may undertake a number of functions within innovation systems including: (1) innovation facilitator – supporting a client firm in its innovation process, for example by bringing a new product to market; (2) innovation carrier – taking a role in transferring existing innovations from one firm or industry to a client firm or industry; and (3) innovation source – playing a role in initiating and developing innovation in client firms (Hauknes, 2000; García-Quevedo and Mas-Verdú, 2008). Accordingly, KIBS are viewed as important components of the knowledge-based economy (Huggins, 2011), complementing the publicly funded knowledge infrastructure and providing a 'second knowledge infrastructure' (Cooke and Memedovic, 2003).

Employment in consumption-based sectors

Another perspective on regional economic development looks at the role of consumption-based sectors in their development. These sectors are typically based around economic activity centring on consumer goods and leisure or recreational activities, i.e. 'experience goods' (Glaeser, 2001; Lorentzen, 2009). Consumption is viewed by some scholars as an 'ongoing process', that is not confined to 'single acts of consumption at formal sites of consumption' (Crewe, 2001: 280). Yet, in spatial terms 'spaces of consumption are also spaces of work,' (Crewe, 2001: 167); thus, the location of one individual's consumption is the location of another individual's employment.

Consumption-based sectors cover a broad range of activities including the retailing of goods and services, restaurants and bars, museums and cinemas. For the purposes of this chapter, this sub-group is further divided into two broad groups: retail services and leisure services. These sectors contribute both directly and indirectly to regional economies; directly in terms of employment and GDP generated, and indirectly through several non-pecuniary mechanisms. For example, consumption-based sectors can play an important role in terms of economic development strategies as regeneration programmes typically 'include an array of consumption spaces from restaurants and tourist zones to museums of art and other cultural fields, gambling casinos, sports stadia and specialised stores' (Zukin, 1998: 832).

Resilience

In light of the economic recession of 2008–2012 many scholars have turned their attention to the concept of resilience (Simmie and Martin, 2010; Martin, 2011). Resilience generally refers to the ability of a regional economy to respond to an event and offer resistance; in short it captures the adaptability of a region to economic shocks rather than just change (Hassink, 2010; Hudson, 2010). Thus, the emphasis on resilience refocuses the debate away from issues of regional competitiveness, an idea that is focused on the long-term economic trajectory of regions (Kitson *et al.*, 2004) to more short-term regional sustainability in terms of maintaining employment levels and standards of living for those resident within a given region at the time of an external shock to the regional system (Bristow, 2010). As such, regional diversity has been demonstrated to be important to resilience, the more varied a region's economy then the more likely it is to survive economic shocks (Swanstrom, 2008). Scholars from the evolutionary economic geography school have further developed these arguments to suggest that it is not variety per se but related variety within the regional economy that is the key to resilience (Frenken *et al.*, 2007; Van Oort *et al.*, 2014). Consequently, regional resilience may be a path dependent process (Hassink, 2010).

Despite the apparent simplicity of the concept there are disagreements within the literature as to what exactly 'resilience' means; is it the ability of a region to return to a previous growth path, i.e. recovery, or is it the ability to a region to adapt and renew its economy (Hassink, 2010)? In light of these debates, this chapter adopts the former approach in that regional resilience is viewed as the ability of a regional economy to return to growth following a period of economic downturn. As such, this chapter is concerned with the role of the service sector in terms of contributing to resilience of European regions. Debates in the literature have identified factors such as a strong regional innovation system as one way of creating a more resilient regional economy (Christopherson *et al.*, 2010; Clark *et al.*, 2010), suggesting that employment within knowledge-based service firms, which have been identified as important components of these systems, may play a role in this. However, given the important role of consumer expenditure in the modern economy little attention has been given to consumption-based sectors and regional resilience within the literature; thus, this chapter begins to address this omission.

Data and method

In order to tackle the research questions set out in the previous section, the analysis utilises data from Eurostat on employment in the service sector, focussing on business services, retail services and leisure and cultural services between 2008–2012. The standard definition of the business services sector was utilised for the purposes of this analysis, focussing on the following activities: computer consultancy, R&D consultancy, and general business services such as accountancy, engineering consultancy, and legal services (Miles, 2005). Retail and leisure and cultural services were more straightforward to define, concentrating on all employment in retail and wholesale for the former and food and drink provision and recreational activities in the case of the latter (Johnston, 2009).

During the years 2008–2012 the European economy experienced a severe recession, followed by a period of economic stagnation initiated by a worldwide banking crisis and marked by low growth or contraction and sovereign debt crises. Therefore, focussing on this time period provides an ideal opportunity to examine the resilience of regional economies with respect to the relative performance of KIBS employment within the region. In this case, a narrow interpretation of resilience is utilised, defined in terms of the change in regional GDP per capita at NUTS 2 level as this indicator captures changes in overall economic activity. While it is acknowledged that resilience is both a broad and contested term (Simmie and Martin, 2010; Martin, 2011), it was felt that for a short exploratory piece of research on the role of the service sector a narrower definition is most useful.

Data for 268 European NUTS 2 regions were obtained; this excludes French overseas territories as these are outside the European continent. The regions were divided into capital city regions in order to assess the performance of these regions separately, defined as any NUTS 2 region containing a capital city. Some capital cities are regions their own right, for example Berlin and Madrid, whereas others are included with their hinterland, for example Warsaw and Rome. All are included for the purposes of the analysis. Finally, London occupies two NUTS 2 regions, Inner London and Outer London, for the purposes of the analysis London is defined as the Inner London NUTS 2 region.

Results

Overview of employment changes

This section first examines employment levels in the three service sub-sectors at the country level in order to explore patterns and variations across countries. Figure 5.1 presents service-sector employment per 10,000 inhabitants for the countries of the European Union. Firstly, this highlights the fact that retail-based employment is the dominant sector out of the three presented, with over 500 people employed in this sector per 10,000 people in almost all countries. Romania (453 jobs per 10,000 inhabitants), Finland (518) and France (522) have the lowest per capita employment in this sector. In contrast Lithuania (793), The Netherlands (877), and Luxembourg have the highest per capita employment in this sector.

Employment in knowledge-based services generally exceeds employment in leisure-based sectors across the countries of the EU. Romania (97), Bulgaria (127) and Poland (142) have the lowest levels of per capita in employment knowledge-based service sectors, whereas Luxembourg (569), the Netherlands (367) and the UK (340) have the highest levels. Finally, in terms of employment in leisure-based services Poland (63), Romania (77) and Slovenia (93) have the lowest levels, and Cyprus (438), Malta (386) and Luxembourg (342) the highest.

The analysis now focuses on changes in service-sector employment across the regions of the EU. In terms of changes in service-sector employment, Table 5.1 highlights differing patterns across the sub-sectors by region. Firstly, in terms of the average size of each sector per region the data suggest that retail employment

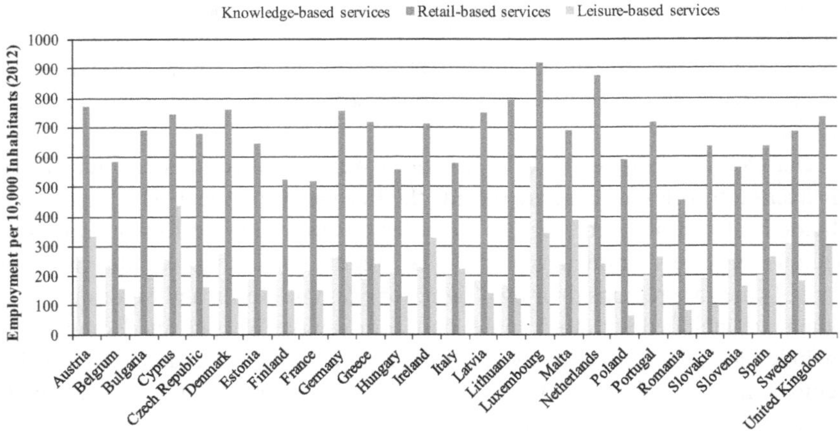

Figure 5.1 Service-sector employment per inhabitant (2012)

outweighs the other two sectors by around 3:1. Thus, on average, in a NUTS 2 region, employment in the retail sector averages 122,000, which is around three times larger than the other sectors which have average employment around 43,000 and 38,000 respectively (see Table 5.1).

In terms of growth rates significant differences are observed; between 2008–12 employment has grown on average by over 6 per cent per region in knowledge-based and leisure-based services whereas employment in retail services shows an average increase of just 0.29 per cent. Thus, while retail employment tends to be larger than the other two sectors within regions, average employment growth within this sector was stagnant between 2008 and 2012. This outcome reflects the decline in consumption expenditure by households across the European Union during this period, where growth averaged 0.2 per cent compared with around 2 per cent per annum for the years preceding the financial crisis (Eurostat, 2013). In contrast, employment growth in the other two sub-sectors appears, on average, to be much more robust. However, when compared to overall employment change, where on average regional employment fell by 3.6 per cent, all three sectors at least exhibit a positive change during the period.

Different types of region display different levels of employment and growth rates on average. The results suggest that capital city regions are outperforming other regions in terms of these indicators. Firstly, in terms of employment, capital city regions, on average, tend to have over twice the level of employment in retail and leisure services and over four times as many jobs in knowledge-based services. This result is, perhaps, unsurprising given that capital cities tend to be larger centres of commerce than other regions. However, in terms of employment growth, these regions, on average, outperform other regions in both the knowledge-based and leisure-based services. The interesting finding is that capital city regions, on average, experienced a decline in retail sector employment, possibly reflecting structural changes taking place within the sector such as increasing levels of e-commerce.

Thus, a number of differences have been identified across the countries and the regions of the European Union in terms of both numbers employed and growth rates across the three sub-sectors of the service sector. The main point to note is

Table 5.1 Regional employment in the European service sector (2008–12)

	All NUTS 2 regions	*Capital city regions*
Mean knowledge-based services employment (2012)	43,596	145,487
Mean retail services employment (2012)	122,579	260,817
Mean leisure services employment (2012)	38,825	88,831
Mean knowledge-based services employment growth (2008–12)	6.49%	7.24%
Mean retail services employment growth (2008–12)	0.29%	–0.47%
Mean leisure services employment growth (2008–12)	6.65%	7.96%

that these differences do not represent a dichotomy between knowledge-based and consumption-based sectors. Instead it appears that knowledge-based services and leisure-based services are broadly similar in terms of growth rates. While the retail services sector is somewhat larger, employment growth has been stagnant during the time period. As such, the evidence suggests that employers in the retail sector have been reluctant or unable to add to their workforces during this period as employment in this sector is generally lower skilled and more casualised. Therefore, firms in this sector may be quick to lay off workers at the start of a recession and slow to re-hire after. In contrast, employment within the other two service sub-sectors may be more resilient as firms seek to protect their human capital as knowledge-based sectors, and to a lesser extent leisure-based sectors, which rely more heavily on such capital to create their services/experiences (Gabe *et al.*, 2012; Donald *et al.*, 2013).

Service-sector employment change and regional resilience

Having examined overall patterns of employment size and change at the regional level, this section now examines the relationship between regional resilience and employment change in each of the three service sub-sectors in order to assess the relationship between employment change in each sector and overall regional economic performance. Firstly, correlations between regional employment within each subsector and change in regional GDP are examined in order to assess broad relationships between the two.

In terms of regional resilience, the analysis firstly found that there was no relationship between the size of knowledge-based employment and overall employment growth in this sector (Pearson Coefficient = 0.032; $p>0.6$) which suggests that it is not the absolute size of employment in this sector that is important, only its dynamism. Figure 5.2 highlights the relationship between the change in employment in knowledge-based services and the change in overall regional GDP. A clear positive relationship exists between the two variables (Pearson coefficient = 0.349; $P<0.001$), suggesting that employment growth in this sector is generally associated with GDP growth. Conversely, those regions with lower levels of employment growth in KIBS exhibit lower, or even negative, levels of GDP growth. This suggests that the presence of knowledge-based sectors may be an important basis for regional resilience. Indeed, higher levels of employment in these sectors may be a signal of a more developed regional innovation system, promoting higher levels of innovation (Cooke and Memedovic, 2003).

Additionally, while there appears to be a positive correlation between employment growth and regional resilience, Figure 5.2 also highlights that within a number of regions employment in this sector did grow but that within the regional economy as a whole GDP growth was negative. Thus, it must be stated that knowledge-based employment is not a panacea, and regional resilience is not guaranteed.

In terms of capital city regions, further analysis highlighted that a similar pattern exists but that the relationship is much stronger (Pearson coefficient =

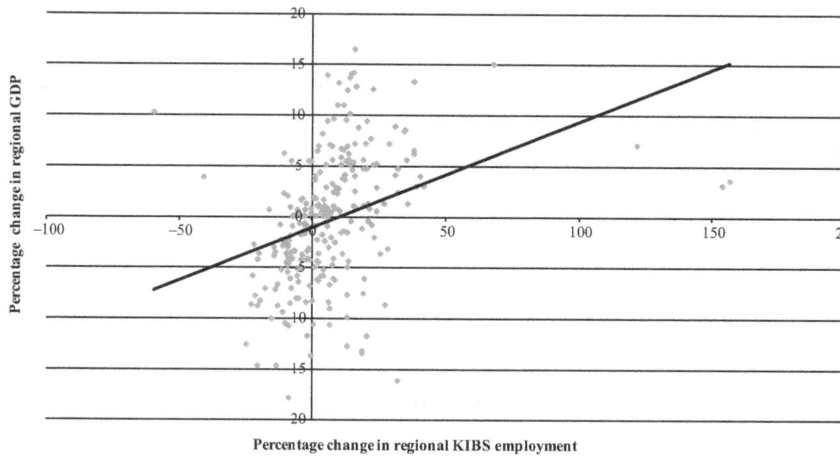

Figure 5.2 Regional employment change in knowledge-based sectors and regional resilience: NUTS 2 regions 2008–2012

0.740; p<0.001); thus, those capital regions where employment in knowledge-based services experienced greater growth performed better in terms of changes in regional GDP. Consequently, it appears that these regions have higher levels of employment in the knowledge-based sectors as well as higher levels of employment growth, which means they are much more reliant upon them for regional growth. As such, the evidence suggests higher levels of knowledge-based employment growth are indicators of more resilient regional economies.

Figure 5.3 presents evidence of a similar relationship between changes in retail employment and regional resilience as to that previously found between knowledge-based employment and resilience. Overall, higher levels of growth in employment in this sector are associated with higher levels of regional growth but the relationship between the two is stronger (Pearson coefficient = 0.436; p<0.001). Again, there are a considerable number of regions where employment in this sector was positive, but overall the regional economy experienced negative growth. As with the previous result for knowledge-based employment in capital city regions, the relationship was found to be stronger for retail employment change and GDP change in capital city regions (Pearson Coefficient = 0.607; p<0.05). Finally, no significant correlation was found between the overall size of employment in the retail sector and changes in regional GDP (Pearson coefficient = 0.034; p>0.500). In addition, Figure 5.3 shows that the distribution of regions is more scattered than in Figure 5.2, suggesting a greater variance in performance among the European NUTS 2 regions in terms of retail employment.

Figure 5.4 shows that correlations between employment growth in the leisure sector and economic resilience again display a similar positive relationship as

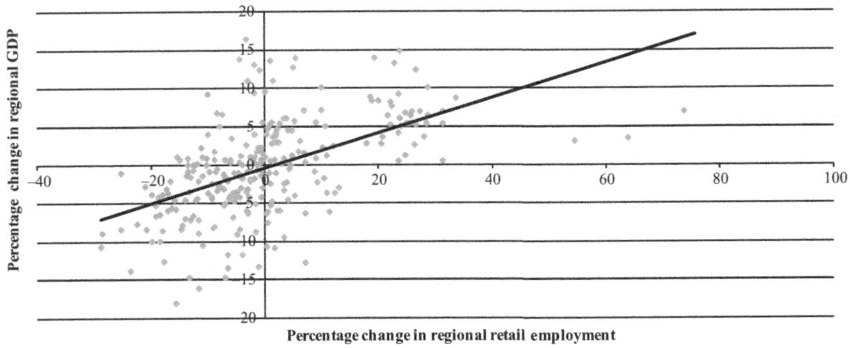

Figure 5.3 Regional employment change in retail sectors and regional resilience: NUTS 2 regions

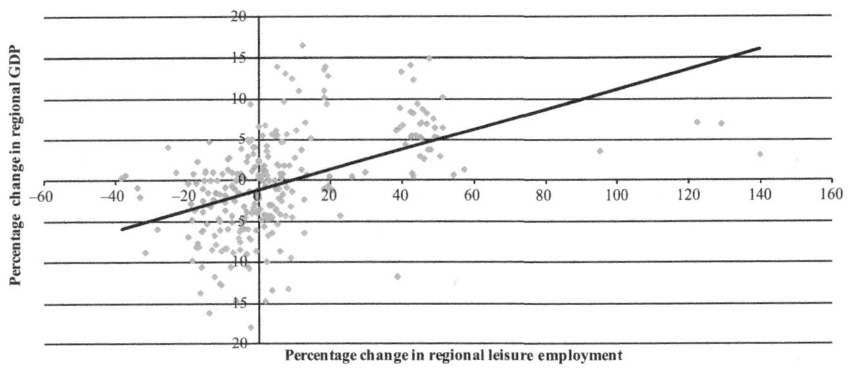

Figure 5.4 Regional employment change in leisure sectors and regional resilience: NUTS 2 regions

observed with the other two sectors (Pearson coefficient = 0.479; $p<0.001$). Again, no significant correlation was observed between the overall size of the leisure sector and regional resilience (Pearson coefficient = –0.037; $p>0.500$), while the relationship was found to be stronger in capital city regions (Pearson coefficient = 0.511, $p<0.05$).

Thus, there appears to be a positive relationship with employment growth in both knowledge-based and consumption-based sectors and regional economic resilience. In the main, where employment in these sectors is growing the region will tend to exhibit a higher rate of overall GDP growth.

Discussion and conclusion

The first conclusion from the analysis is that on average, employment growth per region in business and leisure and cultural services grew by around 6.5 per cent between 2008 and 2012, whereas employment growth per region in retail services averaged 0.3 per cent. Thus, the former two sub-sectors appear to have performed better during the period of recession.

The analysis has highlighted the existence of a positive relationship between regional economic resilience and employment change in the service sector. Thus, a healthy service sector can be seen as a sign of a regional economy that is more capable of withstanding a recessionary shock. Thus, employment growth in service sectors can be seen as a reasonable indicator of the propensity for economic recovery at the regional level. However, there is no evidence to suggest that the overall magnitude of employment in either knowledge-based or consumption-based sectors are responsible for regional resilience as the relative size of employment in each has no bearing on the level of growth in regional GDP. Intead, it is the dynamics of service-sector employment that provides an indicator of resilience.

The analysis also examined whether differences between capital city regions and other regions existed in terms of performance. While capital city regions were found to be larger centres of service-sector employment and benefiting from higher rates of growth, in terms of business and leisure and cultural services, similar patterns were observed with respect to the relationship between employment growth and resilience. Thus, it can be concluded that both types of region are subject to the same dynamics.

The revelatory nature of these findings, i.e. establishing clear relationships between employment growth in the service sector and regional economic performance, suggest further avenues for research. The analysis has highlighted the correlation between service employment growth and regional resilience but it has not established the relative importance of each sector nor any causal link between growth in one sector and regional resilience. This suggests that future work in this area should focus on examining causality in terms of employment in particular sectors and the regions' ability to survive shocks. While some have eschewed this approach, suggesting that examining/treating the concept of resilience as the search for an equilibrium narrows its focus (Pike *et al.*, 2010), it is believed that looking at causality will involve the search for magnitude of effects rather than merely the existence of a relationship.

In terms of policy implications, the findings suggest that service sectors are important components of the regional economy and should be treated as such. In addition, each of the three sub-sectors has been shown to be associated with resilience, not just the more knowledge-based sector of business services. Thus, this provides further evidence that the focus on what has been previously described as the 'high road' of economic development should not be pursued at all costs (Malecki, 2004). Policymakers should endeavour to focus on all branches of the service sector, an approach that would promote variety within the regional economy, which has also been seen by some to be an important driver of growth (Van Oort *et al.*, 2014). Indeed, as the urban form of cities increasingly relies on the retail sector for structure and creation of the public realm it is important that policymakers do not ignore the potential for less knowledge-intensive service sectors in their regional development programmes (Lee, 2012; Wrigley and Dolega, 2011).

References

Bristow, G. (2010). Resilient regions: re-'place'-ing regional competitiveness. *Cambridge Journal of Regions, Economy and Society*, 3(1), pp. 153–67.

Browning, H. and Singlemann, J. (1975). *The Emergence of a Service Society: Demographic and Sociological Aspects of the Sectoral Transformation of the Labor Force in the USA*, Washington D.C.: National Technical Information Service, U.S. Department of Commerce.

Bryson, J.R., Daniels, P.W., Henry, N. and Pollard, J. (2000). *Knowledge, Space, Economy.* London: Routledge.

Chadwick, A., Glasson, J. and Smith, H.L. (2008). Employment Growth in Knowledge-Intensive Business Services in Great Britain during the 1990s – Variations at the Regional and Sub-Regional Level. *Local Economy*, 23(1), pp. 6–18.

Christopherson, S., Michie, J. and Tyler, P. (2010). Regional resilience: theoretical and empirical perspectives. *Cambridge Journal of Regions, Economy and Society*, 3(1), pp. 3–10.

Clark, J., Huang, H.-I. and Walsh, J.P. (2010). A typology of "innovation districts": what it means for regional resilience. *Cambridge Journal of Regions, Economy and Society*, 3(1), pp. 121–37.

Cooke, P. and Memedovic, O. (2003). *Strategies for Regional Innovation Systems*, Vienna: UNIDO.

Corrocher, N. and Cusmano, L. (2014). The "KIBS Engine" of Regional Innovation Systems: Empirical Evidence from European Regions. *Regional Studies*, 48(7), pp. 1212–26.

Crewe, L. (2001). The besieged body: Geographies of retailing and consumption. *Progress in Human Geography*, 25(4), pp. 629–40.

Daniels, P. (1991). *Service Industries: A Geographical Appraisal*, London: Methuen.

Dixon, R. and Thirlwall, A. (1975). A Model of Regional Growth-Rate Differences on Kaldorian Lines. *Oxford Economic Papers*, 27(2), pp. 201–14.

Doloreux, D. and Muller, E. (2007). The key dimensions of Knowledge Intensive Business Services (KIBS) analysis. A decade of evolution. *Working paper on Firms and Regions U1/2007 Frauhofer-Institut fur system und Innovationforschung (ISI), Kahrlsruhe.*

Doloreux, D. and Shearmur, R. (2010). Exploring and comparing innovation patterns across different knowledge intensive business services. *Economics of Innovation and New Technology*, 19(7), pp. 605–25.

Donald, B., Gertler, M.S. and Tyler, P. (2013). Creatives after the crash. *Cambridge Journal of Regions, Economy and Society*, 6(1), pp. 3–21.

Evans, G. (2009). Creative cities, creative spaces and urban policy. *Urban Studies*, 46(5/6), pp. 1003–40.

Freel, M. (2006). Patterns of technological innovation in knowledge intensive business services. *Industry and Innovation*, 13(3), pp. 335–58.

Frenken, K., van Oort, F. and Verburg, T. (2007). Related variety, unrelated variety and regional economic growth. *Regional Studies*, 41(5), pp. 685–97.

Gabe, T., Florida, R. and Mellander, C. (2012). The Creative Class and the crisis. *Cambridge Journal of Regions, Economy and Society*, 6(1), pp. 37–53.

Gallaher, M.P. and Petrusa, J.E. (2006). Innovation in the U.S. Service Sector. *The Journal of Technology Transfer*, 31(6), pp. 611–28.

Gallouj, F. (2002). Innovation in services and the attendant old and new myths. *Journal of Socio-Economics*, 31(1), pp. 137–54.

García-Quevedo, J. and Mas-Verdú, F. (2008). Does only size matter in the use of knowledge intensive services? *Small Business Economics*, 31(2), pp. 137–46.

Gertler, M. (2003). Tacit knowledge and the economic geography of context, or the undefinable tacitness of being (there). *Journal of Economic Geography*, 3(1), pp. 75–99.

Hansen, H. and Winther, L. (2010). The Spatial Division of Talent in City Regions: Location Dynamics of Business Services in Copenhagen. *Tijdschrift voor economische en sociale geografie*, 101(1), pp. 55–72.

Hassink, R. (2010). Regional resilience: a promising concept to explain differences in regional economic adaptability? *Cambridge Journal of Regions, Economy and Society*, 3(1), pp. 45–58.

Hauknes, J. (2000). Dynamic innovation systems: What is the role of services? In: M. Boden and I. Miles (eds), *Services and the Knowledge-based Economy*. London: Routledge.

Den Hertog, P. (2000). Knowledge-intensive business as co-producers of innovation. *International Journal of Innovation Management*, 4(4), pp. 491–528.

Howells, J.R.L. (2002). Tacit knowledge, innovation and economic geography. *Urban Studies*, 39(5–6), pp. 871–84.

Hudson, R. (2005). *Economic Geographies*, London: Sage.

Hudson, R. (2010). Resilient regions in an uncertain world: wishful thinking or a practical reality? *Cambridge Journal of Regions, Economy and Society*, 3(1), pp. 11–25.

Huggins, R. (2011). The Growth of Knowledge-Intensive Business Services: Innovation, Markets and Networks. *European Planning Studies*, 19(8), pp. 1459–80.

Illeris, S. (2007). The nature of services. In: Bryson, J.R. and Daniels, P. (eds) *The Handbook of Service Industries*. Cheltenham: Edward Elgar, pp. 19–33.

Jessop, B. (2000). The state and the contradictions of the knowledge-driven economy. In: Bryson, J.R., Daniels, P.W., Henry, N. and Pollard, J. (eds), *Knowledge Space Economy*. London: Routledge.

Johnston, A. (2009). Which sectors drive regional economic development? Changes in employment in knowledge-based and consumption-based sectors and regional economic performance. *Local Economy*, 24(2), pp. 125–39.

Johnston, A. (2011). The Economic Performance of UK Cities, 1995–2005: Driven by Knowledge-based Sectors or Consumption-based Sectors? *European Planning Studies*, 19(12), pp. 2095–2108.

Jones, J. (2013). *UK Service Industries: Definition, classification and evolution*, London: Office for National Statistics.

Kaldor, N. (1970). The Case for Regional Policies. *Scottish Journal of Political Economy*, 17(3), pp. 337–48.

Kitson, M., Martin, R. and Tyler, P. (2004). Regional Competitiveness: An Elusive yet Key Concept? *Regional Studies*, 38(9), pp. 991–99.

Koch, A. and Stahlecker, T. (2006). Regional innovation systems and the foundation of knowledge intensive business services. A comparative study in Bremen, Munich, and Stuttgart, Germany. *European Planning Studies*, 14(2), pp. 123–46.

Lee, N. (2012). Grim down South? The Determinants of Unemployment Increases in British Cities in the 2008–2009 Recession. *Regional Studies*, 48(11), pp. 1761–78.

Leiponen, A. (2005). Organization of Knowledge and Innovation: The Case of Finnish Business Services. *Industry and Innovation*, 12(2), pp. 185–203.

Lorentzen, A. (2009). Cities in the experience economy. *European Planning Studies*, 17(6), pp. 829–45.

Malecki, E. (2004). Jockeying for Position: What It Means and Why It Matters to Regional Development Policy When Places Compete. *Regional Studies*, 38(9), pp. 1101–1120.

Marato-Sanchez, A. and Cuadrado-Roura, J. (2011). *Analysing the Role of Service Sector on Productivity Growth Across European Regions*, Documentos de Trabajo. Madrid: Universidad de Alcalá.

Martin, R. (2011). Regional economic resilience, hysteresis and recessionary shocks. *Journal of Economic Geography*, 12(1), pp. 1–32.

Massey, D. (1984). *Spatial Divisions of Labour: Social Structures and the Geography of Production*, Basingstoke: Macmillan.

McCarthy, J. (2002). Entertainment-led regeneration: The case of Detroit. *Cities*, 19(2), pp. 105–111.

Miles, I. (2005). Knowledge intensive business services: prospects and policies. *Foresight*, 7(6), pp. 39–63.

Muller, E. and Zenker, A. (2001). Business services as actors of knowledge transformation: the role of KIBS in regional and national innovation systems. *Research Policy*, 30(9), pp. 1501–16.

Van Oort, F., de Geus, S. and Dogaru, T. (2014). Related Variety and Regional Economic Growth in a Cross-Section of European Urban Regions. *European Planning Studies*, pp. 1–18.

Pike, A., Dawley, S. and Tomaney, J. (2010). Resilience, adaptation and adaptability. *Cambridge Journal of Regions, Economy and Society*, 3(1), pp. 59–70.

Porter, M. (2003). The economic performance of regions. *Regional Studies*, 37(6–7), pp. 549–78.

Saxenian, A. (1994). *Regional advantage: culture and competition in Silicon Valley and Route 128*, Cambridge, Massachusetts: Harvard University Press.

Simmie, J. and Martin, R. (2010). The economic resilience of regions: towards an evolutionary approach. *Cambridge Journal of Regions, Economy and Society*, 3(1), pp. 27–43.

Strambach, S. (2008). Knowledge-intensive business services (KIBS) as drivers of multilevel knowledge dynamics. *International Journal of Services Technology and Management*, 10(2/3/4), pp. 152–74.

Swanstrom, T. (2008). Regional Resilience: A Critical Examination of the Ecological Framework. *Institute of Urban & Regional Development*.

Totzer, T. and Gigler, U. (2005). *Managing urban dynamics in old industrial cities – Lessons learned on revitalising inner-city industrial sites in six European case studies*, Paper presented at 2005 European Regional Science Association Conference, Istanbul Turkey.

Uppenberg, K. and Strauss, H. (2010). *Innovation and productivity growth in the EU services sector*, Luxembourg: European Investment Bank.

Wood, P. (2006). Urban development and knowledge-intensive business services: Too many unanswered questions. *Growth and Change*, 37(3), pp. 335–61.

Wrigley, N. and Dolega, L. (2011). Resilience, fragility, and adaptation: new evidence on

the performance of UK high streets during global economic crisis and its policy implications. *Environment and Planning A*, 43(10), pp. 2337–63.
Zukin, S. (1998). Urban lifestyles: Diversity and standardisation in spaces of consumption. *Urban Studies*, 35(5–6), pp. 825–39.

6 The dynamics of service FDI location

A disaggregated industry analysis for Great Britain

Jonathan Jones and Colin Wren

Introduction

Foreign direct investment (FDI) in services has grown dramatically in developed countries over recent decades (UNCTAD, 2015), and in the UK it is now the main form of new inward investment (UKTI, 2013), overtaking that of manufacturing in the late 1990s. While there is substantial research on manufacturing FDI at the subnational level (for example Fallon and Cook, 2010; Pelegrin and Bolancé, 2008), the location of service FDI within countries is relatively under-researched, with most studies at the national level. An exception to this is the recent study by Jones and Wren (2015a), who analyse service FDI location at the UK regional level, but for the service sector as a whole. Since it might be expected that FDI location in different service industries is driven by different factors, then in this chapter a further analysis of these data is undertaken to examine service FDI location at the disaggregated industry level.

Over time, service FDI location is likely to have changed, as UNCTAD (2015) report that foreign investment in service activities such as banking, insurance and transportation was traditionally used to support or complement manufacturing operations, but that more recently it is undertaken in its own right to exploit ownership advantages and generate new business. Market-seeking is a motive for service FDI location (Riedl, 2010), especially in activities that are non-tradable, but resource-motives may be more important for service activities that are tradable, such as access to infrastructure or labour. These motives may play-out differently in different activities, with implications for economic development and policy, but there is limited evidence on this, especially at a subnational level. Kolstad and Villanger (2008) find that FDI is market-seeking in four service activities (that is finance, business activities, transport and trade, comprising wholesale and retail) across 57 countries over 1986–2001, but they also find that manufacturing FDI explains the location of the latter two. This complementarity in manufacturing and service FDI location is also found by Ramasamy and Yeung (2010) and by Bagchi-Sen (1995). However, each of these studies dates back to the 1980s and is at the country level, so that not only do they capture the early-stage nature of service FDI location, but there is no sense of how this has changed over time or for different activities.

In this chapter the location of service industry FDI is explored at the NUTS-2 regional level for Great Britain over the years 1996–2005. It is for four industries: Financial Services, Software and Computer Services, Research and Development and Business Services; which together account for over two-thirds of service FDI in the UK over this period. These years cover the strong growth in UK service FDI, which enables the dynamics of its location to be explored as it has grown. The data are for 'greenfield' investment and expansions to existing foreign-owned plants. The analysis has several strands: it considers the regional distribution of this investment; it analyses its geographical concentration; and it explores the dynamics of service location using Markov transition matrices. Since the NUTS-2 regions are sufficiently disaggregated to identify the more-urbanised areas it enables inferences to be drawn.

Overall, the chapter finds that the location of service investment is disproportionately represented in the more urban and more peripheral areas of Great Britain, but that over time it has converged in location, so that it is more like the regional distribution of service activity. It is evident for each of the four service industries, indicating that it is a general phenomenon as service FDI has grown. The pattern of service FDI location suggests that complementarity to manufacturing FDI is less important, and that factors such as the cost and skills of labour are important, accounting for specialisation in regions of service FDI in different industries. This chapter has implications for economic development, although within the peripheral areas the results suggest that the less-urbanised regions attract relatively little service FDI.

The next section briefly sets out the background to the study. It considers the nature of the FDI data and the four service industries. Section 3 considers how investment in these industries has grown over time and it maps their regional location. Section 4 undertakes the analysis of geographic concentration and location dynamics to consider whether or not it has converged over time to the distribution of UK service activity. Section 5 concludes.

Background to the study

UNCTAD (2015) report that the shift of FDI towards services partly reflects service growth, but also that many services are not tradable and must be consumed where they are produced, so that the principal way of bringing services to foreign markets is through FDI. Further, the countries that have liberalised their service FDI regimes more, have achieved greater inflows. Figure 6.1 plots the share of FDI projects in the 25 European Union countries at 2010 over the period 1997–2010 (excluding Cyprus and Malta), based on the European Investment Monitor. It adopts a nine-fold industrial classification, and over the whole period, 88 per cent of the 35,156 foreign investments are in just two of these: 58 per cent in manufacturing and 30 per cent in financial and business services.[1] Figure 6.1 shows that this service FDI has trended upwards over time in the European Union, reflecting the growth in service activity. It also shows that it accounts for over a third of projects in 2010. However, service FDI is more important to UK

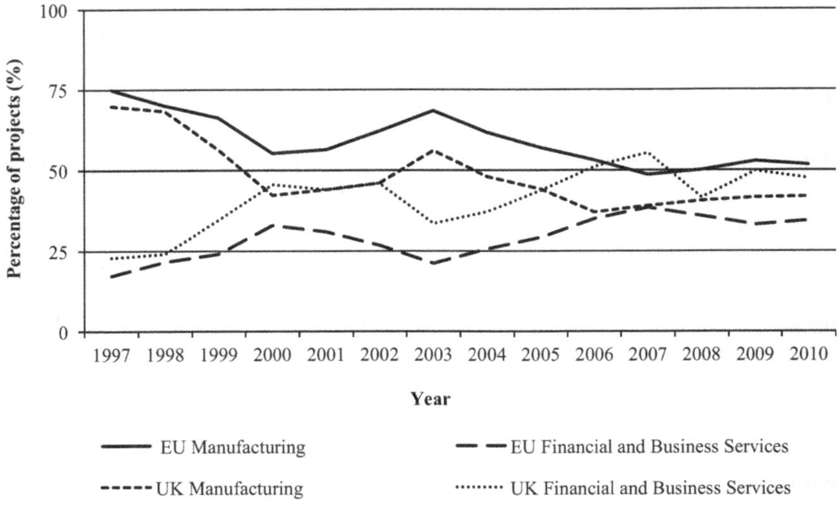

Figure 6.1 Service FDI share in the European Union and UK

Note: EU = 25 European Union countries at 2010, but excluding Cyprus and Malta.

Source: Calculated by the authors from the European Investment Monitor, Ernst and Young (2012).

FDI in each year – the EU's largest recipient – which possibly reflects its greater liberalisation, but also because the UK has the largest financial sector in Europe (UKTI, 2013).

The FDI data used in this study are the same as those used by Jones and Wren (2015a), where further details can be found. These data are project based and were supplied by the main UK national inward investment agency; UK Trade and Investment. They capture the location decision, comprising 'greenfield' FDI and expansions, of which there are a roughly equal number of projects, where the latter represents a substantial upgrading to an existing plant. Compared to mergers and acquisitions (M&A) these are likely to be of greater policy interest at the subnational level since they are gross additions to activity.[2] In total, the data comprise 5,371 foreign investments over 1996–2005, of which 3,000 projects are in services and 2,246 in manufacturing. UK Trade and Investment identifies the location at the NUTS-1 level only, but Jones and Wren (2015a) determine this at the NUTS-2 level, achieving a success rate of 89 per cent overall and 88 per cent for services, which varies little by year or NUTS-1 region. The search often failed to provide sufficient information to determine the NUTS-2 location for London, therefore the NUTS-1 region is used for this, giving 35 regions for Great Britain.[3]

These FDI data are analysed at the regional level in Jones and Wren (2015a), which examines service location in aggregate over 1996–2005. This analysis finds that service FDI tends to locate in the southeast of England, which is due to its greater economic size, but relative to the distribution of service output it tends to locate across Britain and in close proximity to the major urban areas. Over time it has converged in location and so became more evenly spread out relative to the regional distribution of service output. Both the location and convergence of service FDI are spatially independent, so that they are not contingent on the level of FDI in neighbouring regions, and so occur across the British landmass. While service FDI may have located complementary to manufacturing FDI in the late 1990s, the location and convergence of service FDI now occurs at different spatial scales (that is NUTS-2 rather than NUTS-1 regions), so that it does not appear that service FDI locates complementary to manufacturing FDI.

A further study by Jones and Wren (2016) examines the determinants of service FDI location at the NUTS-1 level based on input-output relationships. This uses a similar dataset but over a longer time period, 1985–2007. Compared to manufacturing FDI, it finds that backward non-intermediate effects are relatively more important for service FDI, in the form of labour and non-marketed goods and services (for example infrastructure), but forward intermediate demand from firms is also important (that is producer services). This explains the preference of services for the more urban areas. However, over time, national (rather than regional) consumer demand and regions that serve as a base for exports are more important, where the latter may reflect the formation of the European Union Single Market. Overall, it suggests that market access has become more important for service FDI, while the location of manufacturing FDI is relatively less important. Both these studies are for service FDI as a whole.

FDI location is now explored for individual service industries. Table 6.1 shows service FDI at the 2-digit level using the NACE industrial classification, but with some aggregation/disaggregation for industries with a small/large number of projects.[4] It shows that service FDI is heavily concentrated in a few activities, with divisions 72 to 74 accounting for 1,877 projects. Reflecting this, the analysis is undertaken for four service industries that combined account for about 70 per cent of the service projects over 1996–2005, as follows:

- *Financial Services* (NACE divisions 65 to 71): financial intermediation (banking, leasing and other credit granting), insurance and pension funding, auxiliary intermediation activities (for example fund management), real estate (buying, selling and renting of property) and renting (mainly transport, construction and office equipment).
- *Software and Computer Services* (NACE division 72): computer consultancy (hardware and software) and software (data processing, database, maintenance and other activities).
- *Research and Development* (NACE division 73).
- *Business Services* (NACE division 74): professional services (legal, accounting, marketing, management consultancy and architectural) and other business activities (for example technical testing, advertising, recruitment, cleaning, call centres and miscellaneous).

The groups are each reasonably homogeneous, although each of these may include both high- and low-skill occupations, such as data processing under

Table 6.1 Foreign investment by service industry group

Industry groups	*Number of projects*				*EG index*	
	All years		*1996–2000*	*2001–2005*	*1996–2000*	*2001–2005*
	No.	*(%)*				
Wholesale (50 and 51)	198	(6.6)	90	108	0.022	0.023
Retail (52)	146	(4.9)	40	106	0.009	0.028
Hotels and restaurants (55)	28	(0.9)	10	18	0.463	0.064
Transport and travel (60 to 63)	182	(6.1)	65	117	0.021	0.001
Telecommunications (64)	168	(5.6)	81	87	0.015	0.017
Financial intermediation (65)	148	(4.9)	54	94	0.132	0.064
Insurance and pension funding (66)	13	(0.4)	4	9	–0.085	0.017
Auxiliary financial intermediation (67)	41	(1.4)	17	24	0.006	0.053
Real estate (70)	12	(0.4)	4	8	0.374	0.194
Renting (71)	19	(0.6)	13	6	0.404	–0.068
Computer consultancy (72.1 and 72.2)	665	(22.2)	273	392	0.074	0.041
Computer software (72.3 to 72.6)	426	(14.2)	241	185	0.128	0.045
Research and development (73)	262	(8.7)	48	214	0.024	0.017
Professional business (74.1 and 74.2)	246	(8.2)	93	153	0.066	0.039
Other business activities (74.3 to 74.8)	278	(9.3)	113	165	0.010	0.012
Public administration (75)	14	(0.5)	7	7	0.040	–0.020
Education (80)	30	(1.0)	14	16	0.005	0.095
Health and social work (85)	42	(1.4)	19	23	0.003	0.077
Social and personal services (90 to 99)	82	(2.7)	18	64	0.171	0.087
All services	3,000	(100.0)	1,204	1,796	0.099*	0.041*
All manufacturing	2,246	–	1,234	1,012	0.044*	0.087*
All other	125	–	51	74	0.333*	0.092*
Total	5,371	–	2,489	2,882	0.064*	0.069*

Notes: NACE code (rev. 1.1) in parentheses and EG index by equation (2). * mean value across 19 service, 23 manufacturing and four other industries.

Software and Computer Services or call centres under Business Services, but the industry groups need to be sufficiently large to conduct the transition analysis. They are like Kolstad and Villanger (2008), who report that 86 per cent of global service FDI is in finance, business (including computer activities and R&D), trade and transport. Including retail, wholesale and transport with the above industries gives 2,636 projects for the UK, that is 88 per cent of UK service FDI in Table 6.1. The above four industries continue to be important, accounting for one-third of UK projects in 2012/13 and representing three of the leading recipients of all inward FDI by industry in this year (UKTI, 2013).

The location pattern of service FDI

Over the period 1996–2005, service FDI has grown as a share of total inward UK FDI from 36 per cent in the years 1996–97 to 68 per cent in 2004–05. The time profile of investment for each of the four industry groups is plotted in Figure 6.2. It shows that FDI in each industry has grown over time, but that the number of FDI projects in Software and Computer Services has grown most sharply, accounting for a third of the total service projects over 1996–2005. There is a spike in FDI around the Millennium due to the 'dotcom bubble'. It occurred mainly in computer software (from 19 projects in 1998 to 146 in 2000), but the number of computer consultancy investments also increased (from 43 to 108). Figure 6.2 shows that investment in R&D grew strongly from the year 2002. Business Services include both professional and other business activities and these each have a similar number of investments (Table 6.1).

Figure 6.2 shows that investment in Other Services, outside the above four groups, also grew over the study period, from about 50 to 150 projects a year, of which the major activities are wholesale, retail, transport and telecommunications. Figure 6.2 also shows the number of FDI projects in manufacturing. Like Figure 6.1, this trends downwards over time and by the end of the period it is not that much greater than in Software and Computer Services.

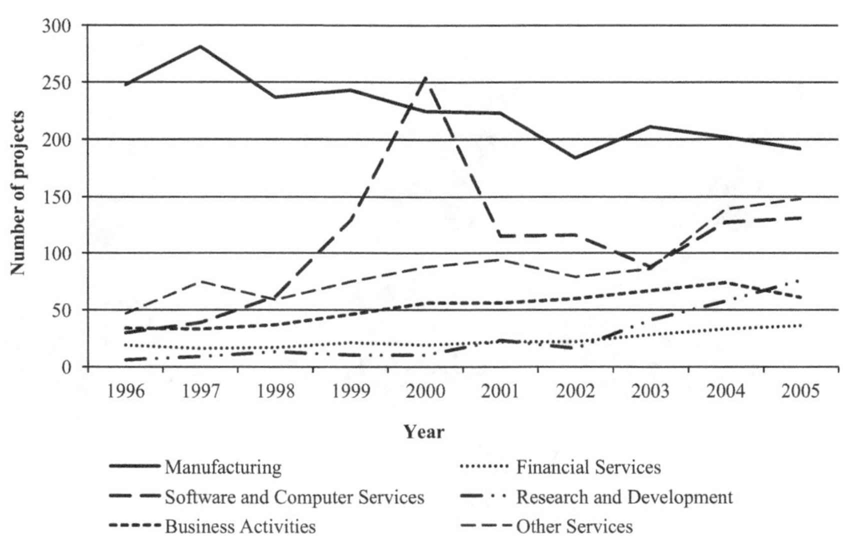

Figure 6.2 UK inward FDI over time by industry group

Source: UK Trade and Investment.

Figure 6.3 shows service FDI location as a whole across the 35 regions of Great Britain over 1996–2005. Each is based on quintile classes, where the class boundaries are given in the key to the respective figures, expressed as decimals. Each class has seven regions, where the regions are identified in Jones and Wren (2015a). Figure 6.3(a) shows the project location of aggregate service FDI and it shows the importance of the more urban areas. There are strong concentrations of service FDI in South-East England, but also in the West Midlands, the North East of England and the West of Scotland. In addition there are concentrations in the regions that include Oxford and Cambridge Universities. Figure 6.3(b) maps the number of jobs associated with the service FDI projects, which shows a similar pattern. The regions in northern Britain continue to be important locations, but South-East England is relatively less attractive, which suggests that on average this tends to get smaller projects.[5]

Figure 6.3(c) maps the aggregate service FDI projects using the location quotient (LQ):

$$LQ_{ir} = \frac{\dfrac{FDI_{ir}}{\sum_r FDI_{ir}}}{\dfrac{GDP_r}{\sum_r GDP_r}} \equiv \frac{\dfrac{FDI_{ir}}{GDP_r}}{\dfrac{\overline{FDI_i}}{\overline{GDP}}}. \tag{1}$$

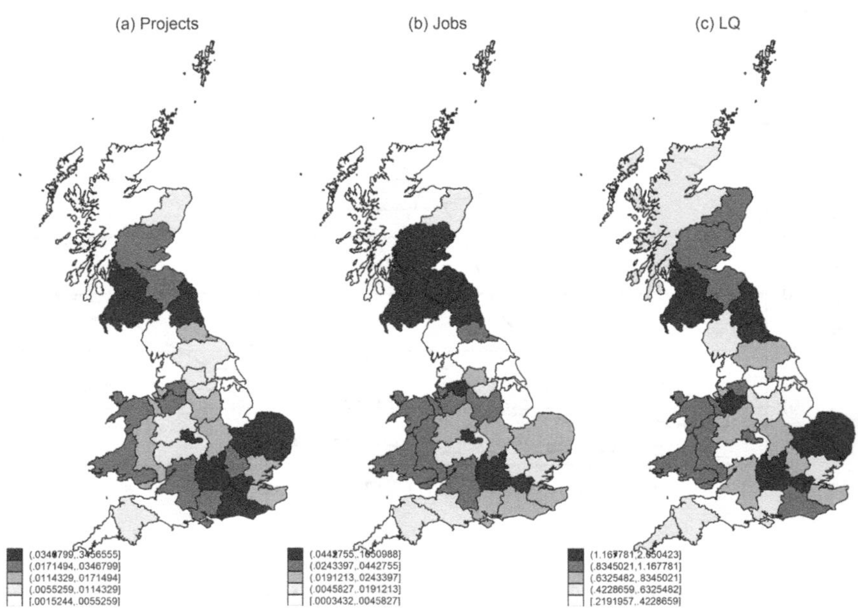

Figure 6.3 Regional location of service FDI, 1996–2005

Sources: Authors' own database and Office for National Statistics. © EuroGeographics for administrative boundaries.

where FDI_{ir} is the number of FDI projects in industry i in region r (= 1, 2, …, 35) and GDP_r is the Gross Domestic Product for services as a whole in r. The LQ may be expressed as the final term in (1), where x-bar is the mean of x across R regions, and this is used to explore regional convergence in FDI location. A value of a location quotient greater [less] than unity indicates that a region gets more [less] FDI in service industry i than may be expected based on the size of its overall service sector. Figure 6.3(c) shows a similar pattern to Figure 6.3(a), but in addition to the more urban areas, Wales and the regions in the northeast of Scotland also attract disproportionately more FDI. There is a cluster of regions in the north of England (Yorkshire, Lancashire and Lincolnshire) that get relatively little FDI.

Figure 6.4 maps the LQs for each of the four industry service groups. Again, these are for the whole period 1996–2005 and relative to total service sector GDP. Overall, they show that different location patterns occur across the British landmass for the four industry groups, which indicates that the patterns exhibited in Figure 6.3 are the outcome of a complex process. In the case of London and South-East England, FDI in Software and Computer Services is overrepresented, as are to some extent Financial Services and R&D, but Business Services is concentrated in the peripheral regions of Wales, North East of England and Scotland. Indeed, Figure 6.4 shows that FDI in each of the industry groups tends to be overrepresented in these regions, with the exception of Software and Computer Services in the North East and Wales. Yorkshire, Lancashire and Lincolnshire do poorly in attracting service FDI in each activity, while the South West of England does poorly in attracting Business Services.

Table 6.2 shows the top-ten NUTS-2 regions in receipt of service FDI, disaggregating these by the four groups. The regions occur across Great Britain and combined they account for 73 per cent of service FDI. It reveals that London, which combines two NUTS-2 regions, receives by far the greatest number of projects, but of which only about 10 per cent are in financial services. This is possibly because most of this investment occurs as M&A. Of much greater importance is FDI in Software and Computer Services, and likewise in other southern regions. In the north, business services are important, particularly around Glasgow and Newcastle, which include the large call centre operations. Further south, Manchester shows a similar pattern of service FDI to Birmingham in the West Midlands, and R&D is concentrated in East Anglia around Cambridge University. Overall, Table 6.2 shows striking patterns of regional specialisation in service FDI location. While UK assistance may be important for locating the larger plants in the peripheral areas, this has tended to focus on manufacturing FDI. Of greater importance is likely to be the availability and cost of different labour skills in these regions.

Geographical concentration and location dynamics

To further explore service FDI location across the regions, Table 6.1 calculates the geographic concentration of FDI location for each industry for two five-year

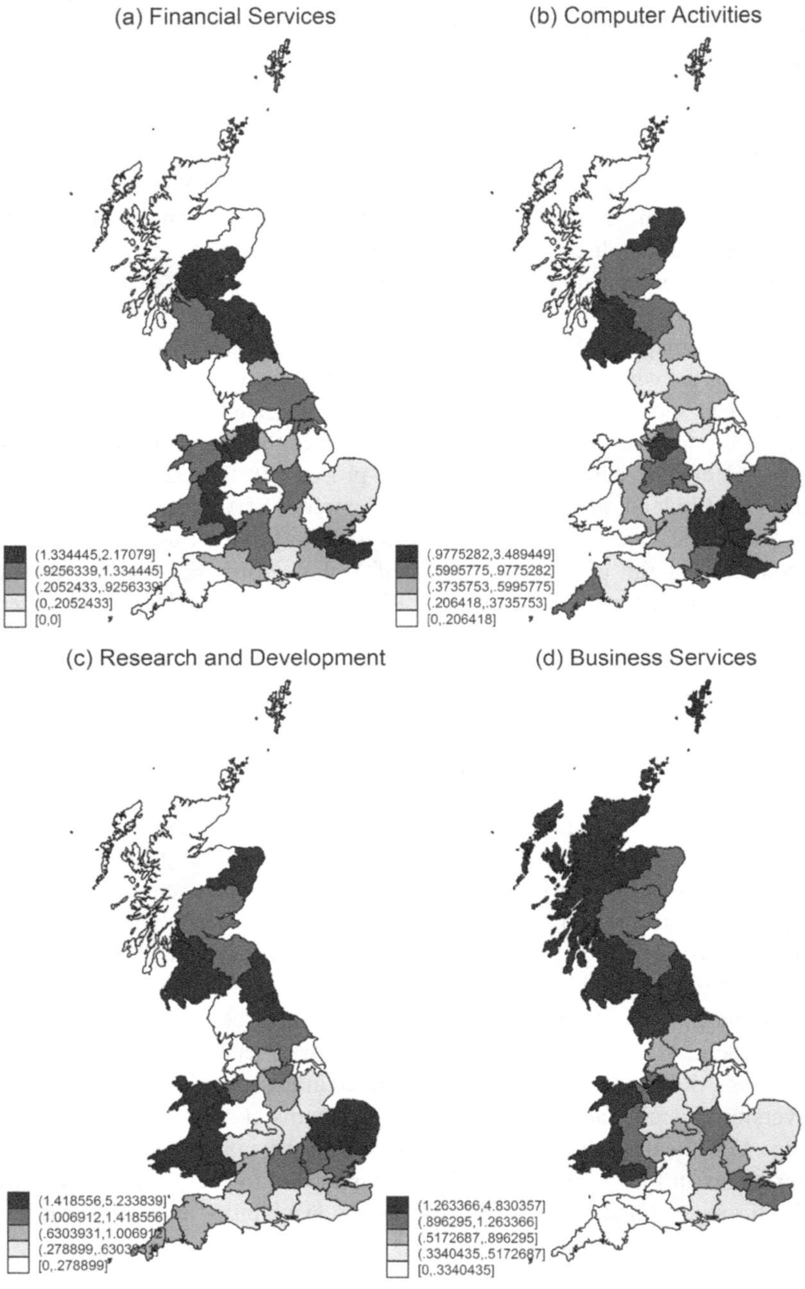

Figure 6.4 Location quotients for service industry groups, 1996–2005

Source: Authors' own database and Office for National Statistics. © EuroGeographics for administrative boundaries.

Table 6.2 Regional specialisation in service FDI location, 1996–2005

Top-ten service FDI recipients:	*Number of service projects*	*Regional project share (%)*			
		FS	*SCS*	*R&D*	*BS*
South:					
London	908	9.7	42.3	5.4	17.3
Bedfordshire and Hertfordshire	249	2.4	66.7	4.8	8.0
Surrey and Sussex	117	7.7	62.4	5.1	8.5
Midlands:					
West Midlands	98	10.2	31.6	3.1	12.2
East Anglia	96	1.0	30.2	39.6	5.2
Gloucestershire, Wiltshire and N. Somerset	79	12.7	26.5	11.4	7.6
North:					
South-West Scotland	114	7.0	28.0	13.1	29.8
Northumberland and Tyne and Wear	98	5.1	10.2	10.2	41.8
Greater Manchester	86	14.0	26.7	2.3	17.4
Eastern Scotland	75	16.0	28.0	10.7	17.3
All regions:	2,625	7.7	36.7	8.9	17.8

Notes: FS = Financial Services; SCS = Software and Computer Services; R&D = Research and Development; and BS = Business Services. NUTS-2 regions identified in Jones and Wren (2015a). FDI for known locations only.

sub-periods. It is based on the Ellison and Glaeser (1997) index (EG), which for industry *i* is:

$$\text{EGIndex}_i = \frac{\dfrac{\sum_r \left(\dfrac{FDI_{ir}}{\sum_r FDI_{ir}} - \dfrac{FDI_r}{\sum_r FDI_r} \right)^2}{1 - \sum_r \left(\dfrac{FDI_r}{\sum_r FDI_r} \right)^2} - \dfrac{1}{n_i}}{1 - \dfrac{1}{n_i}}. \tag{2}$$

This is applied to the number of FDI projects locating across the 35 regions in each industry, and comparison is made with the total number of FDI projects locating over the same period. Broadly, the EG index adjusts the raw geographic concentration index to allow for industrial concentration. In our case, it is the inverse of the number of projects, i.e. $1 / n_i$, which allows for the upwards bias in the raw index due to a small number of projects in some industries. If the industry locates the same as all FDI then raw geographic concentration is zero, so that (2) reduces to $-1 / (n_i - 1)$, which is negative, although close to zero if n_i is large. The EG index increases with geographic concentration, which means that the FDI in the industry becomes more concentrated in its regional location compared to that of all FDI projects.

The EG indices in Table 6.1 show that on average service FDI is more geographically concentrated than manufacturing FDI over 1996–2000, but that the opposite is the case over 2001–05, with the mean EG index across the service industries more than halved. Overall, ten of the 19 industries show a fall in the index between the sub-periods, and while these include a few industries that have only a small number of projects, Table 6.1 shows that it also includes the industries that have a large number of investments, comprising three of the four industry groups, i.e. divisions 72, 73 and 74, as well as financial intermediation (division 65). These account for nearly two-thirds of the investments, and overall they indicate that as service FDI in different activities has grown it has become more spread out in its regional location.

To explore the dynamics of service location a Markov discrete transition probability matrix framework is used. Like the above maps, this involves discretising the distribution of regional FDI location quotients into quintile classes, which reveals important features of the process (Quah, 1996). The analysis is for the location quotients in (1), so that it investigates whether the ratio of the number of projects to service GDP converges towards its mean value across regions over time. The transition matrices for each industry are given in Table 6.3.

Broadly, the approach involves calculating the transition probabilities for each quintile class for the period 1996–2005 as a whole, which shows how the number of investments in each of the five classes (rows) is distributed across the five classes in the subsequent year (columns), so that the probabilities in each row of each transition matrix sum to unity. If this matrix is regular, repeated application of the transition matrix gives the limiting distribution, s_i, which can then be compared with the initial distribution, $s_i(0)$, that is the initial share of regions in each quintile class. If $s_i > s_i(0)$ for the median class(es) but $s_i < s_i(0)$ for other classes with either smaller or larger location quotients, then it is evidence of convergence in FDI location. This is because the regions that initially have high location quotients have smaller quotients over time, and conversely. Both the initial and limiting distributions are shown in Table 6.3.

Part (a) of Table 6.3 gives the results for all service FDI, which is taken from Jones and Wren (2015a). Each class is defined so that initially it has 20 per cent of the regions over the study period, and the limiting distribution shows that service FDI has converged in its location over time. Parts (b) to (e) give the results for the four groups. Since many regions do not get FDI from an industry in a particular year, the class boundaries are set so that Class 1 contains the zero regional FDI shares and the other classes have broadly equal shares. Table 6.3 shows that FDI tends to be concentrated in a few regions for each industry, such that the share of regions with no FDI in a year can be as high as 77.46 per cent for Financial Services. The table shows that there is convergence for Software and Computer Services, Business Services and R&D, but that for Financial Services there is a movement up the classes. This is not divergence, as the regions in Class 1 (that is zero shares) get fewer projects in the limit, and it may just reflect the greater heterogeneity of this industry, but with some activities attracting few projects.

Table 6.3 Transition matrices

Upper bound		*Transition probabilities*					*Initial distribution*	*Limiting distribution*
(a) Service FDI								
	Class	1	2	3	4	5		
0.2208	1	0.397	0.238	0.175	0.095	0.095	0.2000	0.1846
0.5651	2	0.191	0.302	0.286	0.159	0.064	0.2000	0.2099
0.8808	3	0.175	0.222	0.270	0.238	0.095	0.2000	0.2157
1.3721	4	0.079	0.191	0.175	0.365	0.191	0.2000	0.2138
–	5	0.095	0.079	0.159	0.191	0.476	0.2000	0.1760
(b) Financial Services (NACE divisions 65 to 71)								
	Class	1	2	3	4	5		
0.0001	1	0.820	0.049	0.049	0.041	0.041	0.7746	0.7388
1.6160	2	0.529	0.118	0.235	0.000	0.118	0.0539	0.0709
2.3814	3	0.263	0.263	0.158	0.210	0.105	0.0603	0.0747
3.7220	4	0.500	0.056	0.111	0.222	0.111	0.0571	0.0591
–	5	0.823	0.059	0.059	0.000	0.059	0.0539	0.0564
(c) Software and Computer Services (NACE division 72)								
	Class	1	2	3	4	5		
0.0001	1	0.640	0.105	0.072	0.098	0.085	0.4857	0.4268
0.5887	2	0.293	0.244	0.268	0.195	0.000	0.1302	0.1679
1.0740	3	0.150	0.325	0.300	0.125	0.100	0.1270	0.1650
1.7007	4	0.400	0.175	0.200	0.175	0.050	0.1270	0.1342
–	5	0.244	0.049	0.122	0.146	0.439	0.1302	0.1060
(d) Research and Development (NACE division 73)								
	Class	1	2	3	4	5		
0.0001	1	0.750	0.062	0.062	0.049	0.076	0.7111	0.6381
1.3113	2	0.391	0.478	0.130	0.000	0.000	0.0730	0.1508
2.8755	3	0.409	0.409	0.091	0.045	0.045	0.0698	0.0886
5.1887	4	0.435	0.043	0.261	0.217	0.043	0.0730	0.0592
–	5	0.609	0.000	0.087	0.174	0.130	0.0730	0.0633
(e) Business Services (NACE division 74)								
	Class	1	2	3	4	5		
0.0001	1	0.667	0.091	0.091	0.097	0.054	0.5238	0.5128
0.8601	2	0.342	0.368	0.158	0.132	0.000	0.1206	0.1223
1.4808	3	0.297	0.162	0.297	0.108	0.135	0.1175	0.1295
2.4279	4	0.553	0.053	0.105	0.184	0.105	0.1206	0.1267
–	5	0.189	0.027	0.108	0.216	0.460	0.1175	0.1088
*(f) Producer Services**								
	Class	1	2	3	4	5		
0.0001	1	0.480	0.147	0.120	0.133	0.120	0.2381	0.2041
0.5075	2	0.177	0.294	0.373	0.137	0.019	0.1619	0.1721
0.8715	3	0.159	0.222	0.270	0.254	0.095	0.2000	0.2183
1.3836	4	0.079	0.143	0.175	0.333	0.270	0.2000	0.2176
–	5	0.127	0.064	0.175	0.206	0.429	0.2000	0.1879

Note: * OECD definition.

Finally, part (f) of Table 6.3 uses the OECD definition of producer services, comprising business and professional, financial, insurance and real estate. It also suggests convergence, which is no doubt because it is dominated by Business Services (see Table 6.1). However, as Jones and Wren (2015a) find that service FDI locates at a different spatial scale to manufacturing FDI, and that the latter shifts location from North-to-South, then it does not suggest that service FDI is simply following the changing location pattern of manufacturing FDI.

Conclusions

This chapter analyses the location of FDI in service industries across the NUTS-2 regions of Great Britain over the period 1996–2005. This is for 'greenfield' and expansion projects, and it captures the strong growth in service FDI. Four industries are considered, which together account for 70 per cent of FDI projects over this period: Financial Services, Software and Computer Services, R&D and Business Services. It finds that service FDI is highly concentrated, such that ten regions account for around three-quarters of this investment, while in any year many regions do not receive FDI from one or more of these industries. Reflecting this, the regions receiving most service FDI tend to be specialised in the industries in which they receive FDI, so that London gets a high proportion of its service FDI in Software and Computer Services (but less than 10 per cent in Financial Services) and some northern regions in Business Services. In general, service FDI tends to be overrepresented in more urban and more peripheral areas of Great Britain. This is not only the case for each service industry, but it is the case if the economic size of these NUTS-2 regions is controlled for using location quotients.

Over the period 1996–2005, service FDI has grown as a share of total inward UK FDI, from 36 per cent in 1996–1997 to 68 per cent in 2004–2005, such that by the end of this period the number of projects in Software and Computer Services is not much less than in manufacturing. While service FDI is geographically concentrated, as it has grown it has become less concentrated, which is the case for Software and Computer Services, R&D, Business Services and financial intermediation, the main component of Financial Services. It has meant that service FDI has converged in its location as it has grown towards the pattern of UK service activity, but again with the exception of Financial Services. This confirms our earlier result that service FDI has spread out in its location as it has grown, but at the disaggregated industry level.

The implications of the chapter are several. First, the overrepresentation of service FDI in the more peripheral areas could in part compensate for a North-to-South shift that has occurred in UK manufacturing FDI (Jones and Wren, 2012). However, service FDI is also concentrated in regions in the South in industries that are perhaps higher skilled, while within the peripheral areas it locates in or around the more urban areas, which has implications for their overall development. Second, the regional specialisation of service FDI suggests that the availability and cost of labour are important location factors, as well as market access. It suggests that the complementarity to manufacturing FDI is less important over time, which is consistent with earlier work. Notwithstanding this, service FDI is spreading out in its regional location, and in future research it would be useful to explore this by updating the results, and by explaining FDI location at the NUTS-2 level in terms of the specific location factors.

Notes

1 Over the whole period, a further 7.5 per cent is in transport and communications, 2 per cent in retail and hospitality, and the remainder in energy, construction, recreation, education and healthcare, and agriculture.

2 *UK Trade and Industry* also collect data for M&As, but it is often difficult to determine the NUTS-2 location, so that these projects are not included, but in any case the location motives are likely to differ for these.

3 Amendments in 1999 split London into Inner and Outer London NUTS-2 regions. The success rate is applied to the London NUTS-1 region when conducting the regional analysis to ensure comparability.

4 The Nomenclature statistique des activités économiques dans la Communauté européenne (NACE) is the classification of industrial activities used by the European Commission. It was introduced in 1970, revised in 1992 (revision 1.1) and again in 2006 (European Commission, 2008). Details on the classification can be found under the Reference And Management Of Nomenclatures (RAMON) section of the *Eurostat* website.

5 The size distributions of service FDI compared to manufacturing FDI projects are respectively: 1–5 jobs: 25 per cent and 8 per cent; 6–25 jobs: 38 per cent, 26 per cent; 26–200 jobs: 29 per cent, 48 per cent; 201–500 jobs: 6 per cent, 13 per cent; and 501+ jobs: 2 per cent, 5 per cent.

References

Bagchi-Sen, S. (1995), FDI in US Producer Services: A Temporal Analysis of Foreign Direct Investment in the Finance, Insurance and Real Estate Sectors, *Regional Studies*, 29, pp. 159–70.

Ellison, G. and Glaeser, E.L. (1997), Geographic Concentration in US Manufacturing Industries: A Dartboard Approach, *Journal of Political Economy*, 105(5), pp. 889–927.

Ernst and Young (2012), *European Investment Monitor*, Ernst and Young: London.

European Commission (2008), NACE Rev. 2: Statistical Classification of Economic Activities in the European Community, *Methodologies and Working Papers*, Office for Official Publications of the European Communities, Luxembourg.

Fallon, G. and Cook, M. (2010), Exploring the Regional Distribution of Inbound Foreign Direct Investment in the UK in Theory and Practice: Evidence from a Five-Region Study, *Regional Studies*, 44, pp. 37–53.

Jones, J. and Wren, C. (2012), The Location of FDI across British Regions: A Markov Analysis, *Spatial Economic Analysis*, 7(2), pp. 265–86.

Jones, J. and Wren, C. (2015a), Does Service FDI Locate Differently to Manufacturing FDI? A Regional Analysis for Great Britain, *Regional Studies*, forthcoming.

Jones, J. and Wren, C. (2016), Understanding the Location of Foreign Direct Investment, London: Palgrave Macmillan UK.

Kolstad, I. and Villanger, E. (2008), Determinants of FDI in Services, *European Journal of Political Economy*, 24(2), pp. 518–33.

Pelegrin, A. and Bolancé, C. (2008), Regional Foreign Direct Investment in Manufacturing: Do Agglomeration Economies Matter?, *Regional Studies*, 42, pp. 505–22.
Quah, D. (1996), Empirics for Economic Growth and Convergence, *European Economic Review*, 40, pp. 1353–75.
Ramasamy, B. and Yeung, M. (2010), The Determinants of Foreign Direct Investment in Services, *The World Economy*, pp. 573–96.
Riedl, A. (2010), Location Factors of FDI and the Growing Services Economy, *Economics of Transition*, 18, pp. 741–61.
UKTI (2013), *UK Inward Investment Report 2012/13*, UK Trade and Investment, London.
UNCTAD (2015), *World Investment Report 2015: Reforming International Investment Governance*, United Nations, New York and Geneva.

7 Reducing quality uncertainty in the Amsterdam field of restaurants

Pieter Terhorst and Hilal Erkuş-Öztürk

Introduction

All over the world people visit restaurants, order meals, consume them and pay for them every day. Thus actors in the restaurant market seem to coordinate their decentralised actions without any problem. It should not be taken for granted, however, that the restaurant market works well because there is a lot of uncertainty about the quality of restaurants. The restaurant market can only work if quality uncertainty is reduced in one way or another. In the literature two forms of quality uncertainty are found, namely (1) quality uncertainty due to information asymmetry and (2) uncertainty about what constitutes quality. For reasons to be explained below, this paper focuses on the second form of quality uncertainty and discusses how it is reduced in the Amsterdam restaurant market.

According to most economists, quality uncertainty is the result of information asymmetry between sellers and buyers. This is certainly the case in the restaurant industry because the dishes, services, and atmosphere of restaurants are experience goods, by which is meant goods the quality of which can only be known after consumption (Nelson, 1970). In addition, they also are credence goods; those are goods the utility of which is impossible or very difficult for the consumer to ascertain (one cannot be sure about the quality of ingredients used in restaurant meals) (Darby and Karni, 1973). Ever since the seminal paper of Nobel Prize winner Akerlof (1970) economists have argued that quality uncertainty due to information asymmetry between sellers and buyers undermines the functioning of markets because bad products drive good products out of the market ('adverse selection'). Akerlof illustrated adverse selection with an example of the second-hand car market but his arguments can be easily applied to the restaurant market. For instance, in a specific segment of the restaurant market some restaurants offer intentionally inferior quality at given prices. Owners of 'good' restaurants believe their profits are too low at given prices while visitors of 'bad' restaurants believe they pay too much for low quality. As a result, owners of 'good' restaurants will start to leave the market whereas consumers become increasingly reluctant to visit restaurants. In the end, this process leads to a self-destruction of the market. But how then is it possible that a market can survive under conditions of quality uncertainty due to information asymmetry? In other words, how to explain that self-interested rational actors succeed to reduce information asymmetry (an advantageous position for some sellers because it enables opportunistic behaviour!) on behalf of a well-functioning market that is assumed to make everybody better off? Information economists have analysed various ways how and why quality uncertainty due to information asymmetry is reduced, namely through screening, advertisements, guarantees, collective action to create quality labels, state agencies that control quality, and reputation (Macho-Stadler and Pérez Castrillo, 2009: 185–209).

In this paper we approach quality uncertainty from a radically different perspective. Economic sociologists (Aspers, 2009; Beckert, 2009; Bourdieu, 1996; Karpik, 2010) and heterodox economists/economic geographers (Eymard-Duvernay, 1989; Storper and Salais, 1997) argue that the (mainstream) economic analysis of quality uncertainty is limited to so-called 'standard markets' (Aspers, 2009) in which the standards of quality of goods and services are based either on objective criteria, such as in the case of oil or steel, or are so much taken for granted that they become uncontested, 'natural' reference points for quality judgements. But in many other markets, particularly those of cultural products, the quality of goods and services is only partly derived from their objective characteristics. Their value stems largely from symbolic qualities ascribed to goods and services based on interpretation. In so-called 'status markets' (Aspers, 2009) the standards of quality are not objectively given but are contested, differ between classes and lifestyle groups and change over time. Food and drink (wines) offered in restaurants are cultural products because they have a high aesthetic value for the tongue and for the eyes, and have a status-bestowing value too. Classes and lifestyle groups distinguish themselves by food consumption and restaurant visits (Bourdieu, 1984: 175–208; Warde *et al.*, 1999; Warde and Martens, 2000), and many foods and wines served in restaurants are often part of a national, ethnic, religious or regional identity (De Soucey, 2010; Everett and Aitchison, 2008). In addition, the standards of quality change over time. New restaurateurs enter the market and challenge the dominant style of cooking; think of the rise of the *nouvelle cuisine* (Rao *et al.*, 2003) and the diversification of restaurants according to national cuisines. Consumers equally challenge a dominant taste. In the United States, for instance, the dominance of the French *haute cuisine* has been challenged by omnivorous foodies who prefer above all authentic and exotic food (Johnston and Baumann, 2010). Thus in the markets of food, drink, and restaurants the source of quality uncertainty is not so much information asymmetry as well the indeterminacy what constitutes quality (even experienced wine drinkers find it difficult to judge the quality of blindly tasted wines).

If there is so much quality uncertainty in the restaurant market, how then is it possible that it works or exists at all? Not only quality uncertainty due to information asymmetry undermines markets, as Akerlof (1970) argued, but also quality uncertainty that stems from the contingency of the value of goods and services, does. A well-functioning market presupposes some order in which quality uncertainty is reduced (Beckert, 2009). Actors must be able to assess the qualities of goods and services relative to each other and compare them in terms of their value in advance of the actual transaction.

How then is quality uncertainty in the restaurant market that originates from the contingency of the value of goods and services reduced? Much research in economic sociology is done on the processes of qualification of goods and services as an indispensable part of the formation of market value. Concepts like framing (Biggart and Beamish, 2003), status (Aspers, 2009), network (Granovetter, 1995), market devices (Callon *et al.*, 2007), and judgement devices (Karpik, 2010) have been used to investigate how market actors make judgements on the qualities of goods and services offered in the market. Bourdieu (1980, 1984, and 1996) offers a broader theory of how standards of quality in the market or, better, in the field of symbolic goods are established. It integrates the symbolic economy with the material economy, and by conceiving markets as fields it focuses on the structuring forces of market relations and power relations between actors rather than on exchange relations in a narrow sense. In a field perspective, markets are seen as being composed of a set of various actors, such as producers, sellers, experts, journalists, industry associations, regulatory agencies, and consumers. The key actors in the field of restaurants are: restaurateurs, chefs, culinary journalists, publishers of restaurant guides (such as *Michelin Guide* and *Gault Millau*), and consumers. All these actors, who are positioned in a power relation to each other, participate in the social construction of quality of restaurants. Quality is the outcome of a consensus or collective belief of those key actors in the field. That is how quality uncertainty is reduced, at least temporarily because the consensus is challenged continuously.

Does geography matter in the reduction of quality uncertainty? All above-mentioned authors ignore this question altogether. Bourdieu's field is a non-spatial concept but it is plausible to assume that geography matters in how quality uncertainty is reduced. In the field of restaurants, there is on the one hand a tendency to standardisation or McDonaldization (Ritzer, 1998) and on the other hand a process diversification according to ethnic, ecological, vegetarian, and fusion cuisines among others. Standardisation implies predictability and reduction of quality uncertainty at a large geographical scale (a Hamburger of McDonalds in Chicago is exactly the same as in Amsterdam) but diversification, on the other hand, increases quality uncertainty at the local scale as well as, to some extent, at supra-local scale. It is, first, plausible to assume that starters are spin-offs of local incumbent restaurants and imitate local incumbent firms in which they have learned to master the standards of quality. Second, higher-quality restaurants are frequently monitoring each other. It is likely that they do that at the local scale and, to a lesser extent, at supra-local scales. Third, to reduce quality uncertainty of the chefs to be recruited, it is plausible to assume that restaurateurs recruit them via local networks as well as, to a lesser extent, via national and global networks. Fourth, media play an important role in setting the standards of quality and in reducing quality uncertainty. The *Michelin Guide* lists top restaurants only but reviews on sub-top restaurants are predominantly published by national and local media. Thus quality uncertainty is, to a large extent, reduced by national and local media. In addition, it is plausible to assume most reviews on social media such as *Tripadvisor* and *Iens* are written by locals, simply because they are the majority of visitors of local restaurants. *Fifthly*, locals share their experiences of restaurants through predominantly local networks of friends and colleagues who belong to the same class and lifestyle group from which tourists are excluded. It is plausible to assume that tourists reduce their quality uncertainty more by visiting standardised restaurants than locals do. In this paper we explore above-mentioned assumptions on the basis of our empirical research on Amsterdam restaurants.

Amsterdam is a good case-study area because the number of restaurants in the city has strongly increased by 49 per cent in the period 1995–2013 (www.kenniscentrumhoreca.nl) due to a growing demand of locals and tourists. The spending power of locals has increased by a combination of a relatively good performance of the Amsterdam economy and gentrification, and tourism has boomed over the last thirty years. The growth of restaurants has been accompanied by both standardisation and diversification according to various cuisines.

Our paper is organised as follows. In section 2 we discuss Bourdieu's theory of the production and consumption of symbolic goods and apply it to restaurants. In section 3 we present our research methodology, in section 4 our research results, and in section 5 our conclusion.

Bourdieu's economy of symbolic goods applied to restaurants

Bourdieu's (1980 and 1996) analysis of the economy of symbolic goods helps us understand how the restaurant market works. In his analysis of fields of cultural production, he argues that artistic quality is not simply the result of the genius of the artist but the outcome of a collectively produced belief of key actors in a field. Thus artistic or, in our case, culinary quality is the outcome of a (fragile) consensus of those key actors.

By field Bourdieu means a distinctive historically produced social space (for instance the world of science, the arts, or gastronomy) with its own characteristic practices, rules, forms of domination and standards of evaluation. A key characteristic of a field is its objective structure. It is a structured space of positions that is simultaneously a field of *force* operating on those within (similar to soccer in which the action of one player makes all other players change their position) and a field of *struggles* through which actors try to preserve or transform the distribution of resources within it (Bourdieu and Wacquant, 1992: 101–102).

Fields represent the rules of the game that actors implicitly agree to follow but are seldom questioned. Actors in fields act strategically but not according to the theories of rational choice or bounded rationality. Although fields have real and objective structures, actors do not understand them identically. They interpret them through their habitus, that is those deeply imprinted dispositions towards other people, objects and practices in the social field subconsciously acquired from early youth, in education, and in professional life which orient what they think and do. It is a system of internalised dispositions that mediates between the social structure of the field and practical activities in it, being shaped by the former and regulating the latter (Brubaker, 2005: 43). In contrast to rational choice theory, actors follow strategies unconsciously, implicitly and indirectly (see, for unsympathetic and sympathetic criticisms on Bourdieu's 'unconscious strategies', Alexander (1995) and Sayer (1999) respectively).

Actors (in our case chefs, restaurateurs, culinary journalists, and consumers) strategically use their resources – economic capital, cultural capital, social capital, and symbolic capital – to strengthen their position in the field by accumulating one or more of those forms of capital. Each form of capital can be converted in another form. For instance, a good reputation (symbolic capital) eventually leads to a higher income (economic capital). Fields of cultural production are often made up of sub-fields in which the relationship between economic and symbolic criteria of evaluation of goods and services, the habitus of actors, and the amount of different forms of their capital vary. At the one end, there are chefs and restaurateurs who are primarily focused on criteria *outside* the field, that is they try to accumulate economic capital in the short term. At the other end are chefs and restaurateurs who are primarily oriented at field-specific criteria. The more they are oriented at internal, field-specific criteria, the more autonomous a sub-field is. Chefs who show their passion for cooking do not simply produce what generates the highest profits in the short term (if they would do their reputation would be lost) but orient themselves to the standards of judgement in the field. In doing so they follow an 'implicit' or 'unconscious strategy' to get recognition from other chefs in their field by showing them how much they are inspired by them and are simultaneously distinct from them as well as recognition from consumers. The 'strategy' to gain a good reputation should remain hidden because reputation cannot be the result of an explicit strategy. The lack of instrumental calculation is a condition for instrumentally defined success. A good reputation helps actors join field-specific networks (to accumulate social capital) and both eventually lead to a higher income. Because they deny to be driven by economic motives in a narrow sense, it is not surprising that they look down one's nose at chefs who are primarily driven by accumulating economic capital.

But their reputation is not only dependent on recognition from other chefs and consumers but from a lot of other actors who contribute to the production of the collective belief of what good quality is such as culinary journalists, producers of culinary television programmes, and publishers of restaurant guides in gastronomic fields. The status of their judgements is also based on the reputation they have, and they also follow an 'unconscious strategy' to get recognition from others in the field.

Consumers play also an important role in the social construction of quality in the field of restaurants. They are also involved in power struggles to strengthen their position in the field or, more broadly, in different fields. Bourdieu (1984) argues in his *Distinction* that our taste, and indeed all our consumption behaviour, is simultaneously medium and outcome of class relations. Through their consumption behaviour consumers follow also 'unconscious' strategies to strengthen their class position. A successful strategy of distinction requires the aims of distinctive behaviour to be concealed.

The degree of autonomy of sub-fields of restaurants, that is the strength of symbolic criteria of evaluation compared to economic criteria, varies in time and between countries. Ritzer's (1998) analysis of McDonaldization (rationalisation) comes down to an ever growing importance of economic criteria compared to symbolic criteria that seems to be stronger in the US than in most other countries. It implies a reduction of quality uncertainty. On the other hand, a growing importance of high-quality food and a diversification of restaurants according to various cuisines suggest a growing importance of symbolic criteria compared to economic criteria. It implies that the standards of evaluation become ever more fragmented leading to a higher quality uncertainty.

Methodology

Our empirical research is on new restaurants because we assume that the quality uncertainty of both restaurateurs of new restaurants and consumers, that is the 'shared quality uncertainty' (Lupton, 2005), is higher than those of incumbent restaurants and their visitors.

Our research is based on in-depth interviews with 37 restaurateurs who had opened a new restaurant in Amsterdam in the period 2002–2012. The new restaurants were selected from the database Bedrijfsschap Horeca (www.kenniscentrumhoreca.nl) on the basis of a stratified sample according to nine largest ethnic cuisines and thee spatial areas (historic core, South within the ring road, and other neighbourhoods) as well as according to higher-end and lower-end restaurants on the basis of website research. In addition, we interviewed 50 tourists queuing in front of the Van Gogh museum (see for more details Terhorst and Erkuş-Öztürk, 2015). The interviews were done in the summer of 2012 and 2013.

How is quality uncertainty reduced in the Amsterdam field of restaurants?

In this section we aim to answer our research question: 'how is quality uncertainty reduced in the Amsterdam field of restaurants?' by discussing the five hypotheses mentioned in the introduction.

To start a restaurant requires some expertise built up within the sector or in related industries (hotels, pubs) or related professions (catering services, sommeliers) with a short cognitive distance to the restaurant industry, and one should know the Amsterdam market. Otherwise the uncertainty of being successful is too high. In the literature on new firm formation there is overwhelming evidence that most starters set up their firm in the same region they already work (Sorenson, 2003; Stam, 2007). It is, therefore, not surprising that most start-ups are spin-offs from of Amsterdam firms: 23 are spin-offs from Amsterdam restaurants, and eight from other Amsterdam industries. The remaining start-ups are from restaurants elsewhere in the Netherlands.

Reducing quality uncertainty and past carriers of restaurateurs

How starters reduce quality uncertainty depends on the past career of the restaurateurs, that is whether one was a chef before or a manager or waiter. (In our sample 16 restaurants are owned by a restaurateur-chef and 21 by a restaurateur-manager).

Most restaurateurs-chefs have gone through the following standard career. Although their formal training differs (some have been at a specialised cook school, others have no formal training at all), they all have started their working career at a young age as an assistant cook or help in the kitchen. In course of time, they have developed their skills by learning-by- doing in a master-apprenticeship relation. After having mastered the basic skills, the more talented cooks move from one restaurant to another with the same ethnic kitchen (preferably a more prestigious one) to learn from chefs with a higher reputation (there is a lot of job hopping between restaurants of the same ethnic kitchen in Amsterdam), and, finally, to become a chef by themselves. Under that condition, you can exactly cook and run your own business in the way you want. Thus in their career restaurateurs have acquired the habitus, skills, and tacit knowledge of a chef through which quality uncertainty is strongly reduced.

Obviously, not all restaurateurs-chefs have followed the standard career described above. Their carriers vary with their talent, the reputation of the chefs who were their master and the restaurants they were employed, and, last but not least, the type of kitchen they have learned to master. In the kitchen of a standardised restaurant you do not learn much as a result of which it is difficult to find a job in a better restaurant. In addition, cooks and chefs who have been trained in a specific ethnic kitchen or style of cooking, cannot easily and do not switch to other ones. Given the fact that specific cooking skills and a specific culinary habitus cannot be easily changed, it is not surprising that *all* restaurateurs-chefs started up their new restaurant in precisely the same type of kitchen they have learned to master. Even the industry outsider, a former cook in a home for the elderly, cooks in his new 'traditionally Dutch' restaurant (with dishes such as pea soup with sausage, and stuffed beef with red cabbage and mashed potatoes) in the same way as before.

The professional background of restaurateurs-managers differs from that of restaurateurs-chefs. Only four among a total of 21 restaurateurs-managers worked as a cook or chef in a former restaurant while the rest was a manager or waiter, and there are more industrial (7) and geographical (3) outsiders among them. Finally, because their cooking skills and culinary specific habitus are less developed than those of restaurateurs-chefs, restaurateurs-managers (who are industry insiders) are more flexible in switching to a new type of restaurant they have no experience with. For instance, two switched from the declining Chinese-Indonesian market to the more profitable Japanese market, one from a Chinese restaurant outside Amsterdam to a fusion kitchen in Amsterdam, and one Dutch owner of a pizzeria, a simple Dutch-French restaurant, and a few pubs has followed a diversification strategy without being motivated by any culinary motive at all.

Industry outsiders are by definition more flexible than industry insiders in deciding what type of restaurant they start. However, to reduce quality uncertainty one should know the local market as most industry outsiders did (it is no coincidence that no industry outsider came from elsewhere). Four industry outsiders, frequently dining out, decided to start a restaurant on the basis of their experience as a consumer. For instance, a former owner of a discotheque realised that he was missing a typical French bistro in Amsterdam and decided to enter the field (thereafter he opened a French top restaurant), one restaurateur (who worked in a pub before) opened a Spanish tapas restaurant because she likes tapas so much, and two Turkish restaurateurs (a former textile trader and an owner of a catering firm) decided to start their own restaurant because of similar negative experiences with visiting Amsterdam restaurants (as Muslims they felt uncomfortable when people were gazing at their head-scarfed wives and were drinking alcohol). The remaining industry outsiders (except two) have become full insiders in a gradual process, they first stepped with one foot in the industry but kept their former job and only later stepped in with their second foot. Thus almost all restaurateurs-managers/industry outsiders became insiders on the basis of their consumer experience or in a gradual process.

At first sight, it seems that quality uncertainty is higher for restaurateurs-managers than for restaurateurs-chefs because the former have little experience with cooking which is the core business of every restaurant. But this is not true because restaurateurs-managers reduce quality uncertainty much more by opening lower-quality, standardised restaurants (their quality uncertainty is reduced by standardisation) than restaurateurs-chefs do. On the basis of reviews of culinary journalists, our study of menus and, in a number of cases, our own dining experience, we distinguished five quality levels of restaurants: ‘high’, ‘craft high’, ‘craft’, ‘basic craft’, and ‘basic’. High-quality restaurants serve innovative dishes of top quality (the best French, Italian, and Chinese-French restaurants in town) while craft-quality restaurants show their craftsmanship in their dishes but are nevertheless somewhat standard (a few Chinese and Thai restaurants). Basic restaurants are, for instance, Italian restaurants that serve industrially made lasagne and cannelloni, and Argentinean restaurants that mainly serve steaks, French fries and salads. The findings of a similar research (Terhorst and Erkuş-Öztürk, 2015), showed that ‘high-quality’- and ‘craft high’- quality restaurants are associated with restaurateurs-chefs. In short, restaurateurs-chefs start more restaurants in the higher ends of the market than restaurateurs-managers, simply because their habitus and culinary skills are different.

Monitoring

In the literature on the production of symbolic goods it is stressed that actors who are focused on the dominant standards of the field monitor each other, are inspired by each other but try nevertheless to distinguish themselves from others, and try to get recognition from each other. The same applies to chefs and restaurateurs of high-end (French) restaurants (Olders, 2013; Ottenbacher and

Harrington, 2007; Rao *et al.*, 2003), on condition at least they belong to the same type of kitchen. Restaurateurs and chefs specialised in one type of kitchen are generally hardly interested in other kitchens, which is why the field of restaurants is highly fragmented.

In our sample, there is only one top restaurateur-chef of a French restaurant who is a member of a close friendship network with other Amsterdam top restaurateurs-chefs, including a few well-known television-chefs. All other chefs who want to learn from others get their inspiration in other ways, more often than not from chefs and restaurateurs outside Amsterdam. First, there are culinary competitions, in which dishes are compared and selected for a prize. For instance, the election of the Dutch representative for the prestigious Paul Bocuse d'Or competition in Lyon and the competition organised by the Thai Restaurant Association. Second, the chefs of one restaurant followed a summer course given by the famous chef René Redzepi of Noma in Copenhagen. Third, and most importantly, restaurateurs and chefs monitor other restaurants in Amsterdam and elsewhere by visiting them. Not all restaurateurs and chefs monitor other restaurants. Chefs and restaurateurs of lower-end restaurants hardly ever do. They have no reason to do so because their dishes are standardised. This is confirmed by the results of a correspondence analysis we made in our project (Terhorst and Erkuş-Öztürk, 2015). 'Craft', and particularly, 'craft high-quality' and 'high-quality' restaurants monitor other restaurants much more than 'basic' and 'basic craft' restaurants.

Reducing quality uncertainty in finding good chefs

When starting up a restaurant, restaurateurs-managers have to find a good chef to serve dishes of good quality. Quality uncertainty on the output side is accompanied by quality uncertainty on the input side. This obviously applies much more to specialised, higher-quality restaurants than in standardised, lower-quality restaurants because cooking in the latter is relatively simple. Because official diplomas do not say very much about an applicant's quality, restaurateurs-managers are dependent on their social networks to find a chef of good reputation. This is the best way to reduce quality uncertainty about their chefs to be recruited. Almost all restaurateurs-managers, including industry outsiders, have found them in this way (only one did via a specialised agency). It is worth mentioning that all restaurateurs only know other people in the field (most of them in Amsterdam) who have a similar position and reputation. Restaurateurs and chefs of high reputation do not know and have no interest to know restaurateurs and chefs of low reputation. Thus there is status hierarchy in social networks.

For Asian, Indian, and Turkish restaurateurs it is particularly difficult to find good chefs due to strict immigration rules (that have released somewhat for Asian cooks after the finishing of this research). According to those rules, restaurateurs have to make serious attempts to find a chef in the EU during six months. It is not until after that period they are allowed to recruit a chef from elsewhere. Most ethnic restaurateurs-managers finally succeeded to find chefs via their personal networks with their mother country.

Media

Media are very important for new restaurants and new restaurants are important for the media. Good reviews in national and local media, and being listed as a good restaurant in, among others, the *Michelin Guide*, *GaultMillau* guide or the Dutch culinary journal *Lekker* reduces quality uncertainty for consumers, boosts up the number visitors and turnovers, and help starters to overcome the first year that is always difficult for every new firm.

New restaurants, on the other hand, are also important for the media. Culinary journalists of newspapers and specialised culinary journals seldom write about incumbent restaurants; it is their profession to write about news, which are *new* restaurants and *new* trends in the world of food. In addition, specialised culinary journals need new recipes of good dishes. A few restaurateurs-chefs in our example wrote recipes for the culinary journal *Delicious* in exchange for a review of their restaurant.

Just as there are differences in status and reputation of restaurateurs and chefs, there are differences in status and reputation of culinary journalists. The *Michelin Guide* is generally seen as most prestigious guide and reviews top restaurants only. *GaultMillau*, and national and local media are second in the hierarchy of reputation and predominantly on sub-top restaurants. In Amsterdam, the late Johannes van Dam who wrote for the local newspaper *Het Parool*, was *the* culinary authority of the city on sub-top restaurants. If he wrote a good review, you would be fully booked during the first months after opening. At the bottom of the hierarchy are the internet sites like *Tripadvisor* and the Dutch *Iens* in which restaurants are anonymously reviewed. The reader does not know what the amount of culinary specific cultural capital and reputation of the reviewers is.

The media strengthen the differences between lower- and higher-end restaurants. The higher the prestige of a medium (*Michelin Guide*), the higher the quality and reputation of a restaurant is. Culinary journalists never review new low-quality restaurants because they are not interested in it and, more importantly, it would reduce their own reputation. On internet sites such as *Tripadvisor* and *Iens* all types of restaurants in our sample, including the 'basic' and 'basic craft' ones, have been reviewed. It is likely that locals write more reviews on internet on local restaurants than tourists do, simply because they are, in general, the majority of visitors of local restaurants. It would mean that the social construction of quality standards on internet is predominantly made at the local level. As far as we know, no research has been done on that question (nor whether reviewers with a lot of culinary specific cultural capital write predominantly on higher-end restaurants on internet). Do reviews on internet imply a 'democratisation' of how quality standards are set and undermine the reputation of professional reviewers in the written media? It is probably not. Davis' (2012: 190–228) careful research on restaurant reviews shows that reviews on internet, to a large extent, reflect the opinion (and even the formulations) of reviews in the written media. It is in accordance with Bourdieu's theory.

Difference between locals and tourists

Consumers of restaurants also contribute to their social construction of quality. Locals who go dine out regularly build up a lot of experiences on quality of restaurants that are shared with friends, family members and colleagues who belong to the same national or local food culture, class or lifestyle group. Although personal networks can and often do extend at a large spatial scale, it is plausible to assume that their density is highest at the local level. In addition, locals have easily access to local and national media. Although both locals and tourists have access to internet and social media that give a lot of information on restaurants and help reduce quality uncertainty to both, the position of tourists in the field, however, differs from that of locals. *Firstly*, tourists are, generally, not part of local personal networks. *Secondly*, they are excluded from reviews in national and local media. *Thirdly*, the 'strategies' of distinction followed by elites vary from country to country (Daloz, 2010). 'Strategies' that are successfully followed at home do not work abroad. Thus there is a disjunction between the tourists' habitus and field in the place of destination.

Due to these differences, the quality uncertainty of tourists is higher than that of locals. That is why tourists, including the highly educated ones, are inclined to reduce their quality uncertainty by visiting standardised, cheap restaurants. Terhorst and Erkuş-Öztürk (2015) show that almost all highly educated tourists in Amsterdam, most of whom see themselves as a foodie, visit 'basic' and 'basic craft' restaurants. They did not know what value for money in high-quality restaurants in Amsterdam you get (prices in Dutch restaurants are relatively high compared to other European countries). Surprisingly, only a few tourists used internet in choosing their restaurant; most of them found their restaurant by walking around. This is in accordance with the market/locational strategy of restaurateurs. Given their resources, restaurateurs of new restaurants know quite well what type of restaurant they can set up and what type not. Restaurateurs with a lot of culinary specific cultural capital, by contrast, even do not consider the option to open a restaurant in the lower ends of the market. They know their strengths, are focused on the higher ends of the market and try to get a good reputation by serving high-quality dishes.

The above-described differences largely explain their difference in orientation on the tourism market and their locational strategies. 'Basic' and 'basic craft' restaurants are predominantly located in passers-by streets and are strongly dependent on tourists whereas 'high-quality' and 'craft high-quality' restaurants are located in side streets and 'hidden places' in the city and predominantly visited by locals (see for a detailed analysis Terhorst and Erkuş-Öztürk, 2015).

Conclusion

In this paper we distinguished two forms of quality uncertainty, namely quality uncertainty due to information asymmetry and uncertainty about what constitutes quality. Given the symbolic character of food and restaurants, we discussed the second form of quality uncertainty and argued on the basis of Bourdieu's work that quality uncertainty is reduced by a (fragile) consensus of key actors in the field of restaurants. Does geography matter in reducing quality uncertainty in the Amsterdam field of restaurants? Bourdieu's field is a non-spatial concept but we brought in geography in it and showed that quality standards are largely set locally as well as nationally, and, to a lesser extent, globally.

Most starting restaurateurs have learned the required quality standards in the restaurants they worked before (most in Amsterdam and a few elsewhere), and all industry outsiders are from Amsterdam. The latter had learned to know the Amsterdam market as a consumer and/or a part-time restaurateur. Restaurateurs-chefs have a larger impact on the dominant quality standards than restaurateurs-managers because the latter simply lack the required cooking skills, and are much more oriented at the higher-end market than the latter. To keep up with the required quality standards, restaurateurs of specialised, higher-end restaurants monitor each other more than those of lower-end, standardised restaurants for whom the need to do so is less. They monitor each other predominantly at the local and national scale and, to a lesser extent, at the international scale.

Because quality uncertainty on the output side is accompanied by quality uncertainty on the input side, chefs in particular, restaurateurs-managers are very dependent on personal networks to find good chefs. Given the high concentration of restaurants in Amsterdam and the large number of local spin-offs, it is not surprising that restaurateurs-managers find most of their chefs through personal networks in the Amsterdam region and, to a lesser extent, via-via in the Netherlands. Asian, Indian, and Turkish restaurateurs find almost all of their chefs through personal networks with their home country.

Media play an important role in setting quality standards and in reducing quality uncertainty. The position of culinary journalists in de field differs according to their status and reputation. The media strengthen the differences between lower- and higher-end restaurants. The higher the prestige of a medium (*Michelin Guide*), the higher the quality and reputation of a restaurant is. National and local media generally review sub-top restaurants whereas on internet sites all types of restaurants are reviewed. It is unclear yet whether the majority reviewers on local restaurants on internet are locals or tourists. There are indications that reviews on internet sites have not undermined the status and prestige of professional reviewers in the written media.

The quality uncertainty of locals is lower than of tourists. Locals have better access to local and national media and, maybe more importantly, they exchange knowledge about local restaurants via their personal networks of friends, family members, and colleagues who belong to the same class and lifestyle group. Our research shows that (highly educated) tourists reduce their quality uncertainty more by visiting lower-end, standardised restaurants than locals do.

References

Akerlof, G.A. (1970). The Market for "Lemons": Quality Uncertainty and the Market Mechanism. *Quarterly Journal of Economics,* 84(3), pp. 488–500.

Alexander, J.C. (1995). *Fin de Siècle Theory. Relativity and Reduction, and the Problem of Reason*. London/New York: Verso.

Aspers, P. (2009). Knowledge and Value in Markets. *Theory and Society*, 38(2), pp. 111–31.

Beckert, J. (2009). The Social Order of Markets. *Theory and Society*, 38(3), pp. 245–69.

Bedrijfschap Horeca, www.kenniscentrumhoreca.nl [accessed on 17 March 2015].

Biggart, W.N. and Beamish, Th.D. (2003). The Economic Sociology of Conventions. Habit Custom, Practice, and Routine in Market Order. *Annual Review of Sociology*, 29(1), pp. 443–64

Bourdieu, P. (1980). The Production of Belief: Contribution to an Economy of Symbolic Goods. *Media, Culture, and Society*, 2(3), pp. 261–93.

Bourdieu, P. (1984), *Distinction*. Cambridge: Polity Press.

Bourdieu, P. (1996). *The Rules of the Art. Genesis and Structure of the Literary Field*. Redwood City, CA: Stanford University Press.

Bourdieu, P. and Wacquant, L.J. (1992). *An Invitation to Reflexive Sociology*. Oxford: Polity Press.

Brubaker, R. (2005). Rethinking Classical Theory: the Sociological Vision of Pierre Bourdieu. In D.L. Schwartz and V.L. Zolberg (eds), *After Bourdieu. Influence, Critique, Elaboration*, pp. 25–64. New York: Kluwer.

Callon, M., Méadel, C. and Rabeharisoa, V. (2007). *Market Devices*. Oxford: Blackwell.

Daloz, J.P. (2010). *The Sociology of Elite Distinction. From Theoretical to Comparative Perspectives*. London: Palgrave Macmillan.

Darby, M. and Karni, E. (1973). Free Competition and the Optimal Amount of Fraud. *Journal of Law and Economics*, 16(1), pp. 67–88.

Davis, M. (2012). A Taste for New York: Restaurant Reviews, Food Discourse, and the Field of Gastronomy in America. *AAWE Working Paper 106*, www.wine-economics.org [accessed on 12 May 2015].

De Soucey, M. (2010). Gastronationalism: Food Traditions and Authenticity Politics in the European Union. *American Sociological Review*, 75(3), pp. 432–55.

Everett, S. and Aitchison, C. (2008). The Role of Food Tourism in Sustaining Regional Identity: A Case Study of Cornwall, South West England. *Journal of Sustainable Tourism*, 16(2), pp. 150–167.

Eymard-Duvernay, F. (1989). Conventions de Qualité et Formes de Coordination: l'Economie des Conventions. *Revue Economique*, 40(2), pp. 329–59.

Granovetter, M. (1995). Getting a Job: *A Study of Contacts and Carriers*, 2nd edn, Cambridge, MA: Harvard University Press.

Johnston, J. and Baumann, S. (2010). Foodies. Democracy and Distinction in the Gourmet Foodscape. New York: Routledge.

Karpik, L. (2010). *Valuing the Unique: the Economics of Singularities*. Princeton: Princeton University Press.

Macho-Stadler, I. and David Pérez, J. (2009). *An Introduction of Economics of Information*, 2nd edn, Oxford: Oxford University Press.

Nelson, P. (1970). Information and Consumer Behaviour. *Journal of Political Economy*, 78(2), pp. 311–29.

Olders, P. (2013). *Continuity and Change. Essays on Economic Geography and Good Food*. Ph.D. Uppsala University.

Ottenbacher, M. and Harrington, R. (2007). The Innovation Process of Michelin-starred Chefs. *International Journal of Contemporary Hospitality Management*, 19(6), pp. 444–60.

Rao, H., Monin, P. and Durand, R. (2003). Institutional Change in Toque Ville: Nouvelle Cuisine as an Identity Movement in French Gastronomy. *American Journal of Sociology*, 108(4), pp. 795–843.

Ritzer, G. (1998). *The McDonaldization Thesis*. London: Sage.

Sayer, A. (1999). Bourdieu, Smith and Disinterested Judgement. *Sociology Review*. 47(3), pp. 403–31.

Sorenson, O. (2003). Social Networks and Industrial Geography. *Journal of Evolutionary Economics*, 13, pp. 513–27.

Stam, E. (2007). Why Butterflies Don't Leave. Locational Behaviour of Entrepreneurial Firms. *Economic Geography*, 83(1), pp. 27–50.

Storper, M. and Salais, R. (1997). *Worlds of Production: the Action Frameworks of the Economy*. Cambridge, MA: Harvard University Press.

Terhorst, P. and Erkuş-Öztürk, H. (2015). Urban Tourism and Spatial Segmentation in the Field of Restaurants: The Case of Amsterdam. *International Journal of Culture and Hospitality Research*, 9(2), pp. 85–102.

Warde, A. and Martens, L. (2000). *Eating Out: Social Differentiation, Consumption and Pleasure.* Cambridge: Cambridge University Press.

Warde, A., Martens, L. and Olsen, W. (1999). Consumption and the Problem of Variety: Cultural Omnivorousness, Social Distinction and Dining Out. *Sociology*, 33(1), pp. 105–27.

8 Offshoring, outsourcing and services FDI in Europe's 'old periphery'

Probing the experience of Northern Ireland

Mike Crone

Introduction

This chapter is concerned with foreign direct investment (FDI) related to service offshoring and outsourcing and its implications for host economies. Service offshoring involves the 'functional de-coupling' or 'unbundling' and geographical relocation by a firm of certain 'white-collar' service activities, processes, or tasks from one country to another country (Doh, *et al.,* 2008; UNCTAD, 2004). This can be achieved via two main 'governance modes' – dubbed 'captive offshoring' or 'offshore outsourcing' (Sako, 2005; UNCTAD, 2004) – as depicted in Figure 8.1 (Panels A1, B1 and B2). Whilst service offshoring does not always involve FDI, inward-investment projects can flow into host economies in two main scenarios: (1) When a foreign firm establishes a wholly (or partly) owned 'captive centre' in the host economy, which supplies services to the parent company's operations in its home country and/or its other foreign locations; (2) When foreign-owned third-party business process outsourcing (TP-BPO) vendors locate a service delivery centre in the host economy to supply services to clients or consumers in the local market and/or to 'export' services to clients or consumers in foreign markets. The notable global upsurge in such investments over recent decades has made service offshoring an important topic of interest to scholars and policy-makers working in the fields of inward investment and regional development (Hardy *et al.,* 2011; UNCTAD, 2004).

To date, the relatively small but growing literature on offshore services, FDI and regional development issues has primarily focused on host locations in emerging economies, such as India, the Philippines, South Africa and those in Central and Eastern Europe (Benner, 2006; Dossani and Kenney, 2007; Hardy *et al.,* 2011; Kleibert, 2014, 2015; Micek *et al.,* 2011; Pandy and Rogerson, 2012; Sass and Fifekova, 2011). However, this type of FDI has also been attracted to various peripheral, lagging and 'old industrial' regions of advanced economies in Western Europe and North America – a fact barely addressed by the recent academic literature. This raises some interesting questions about the particular type of services projects that have been attracted to such regions and

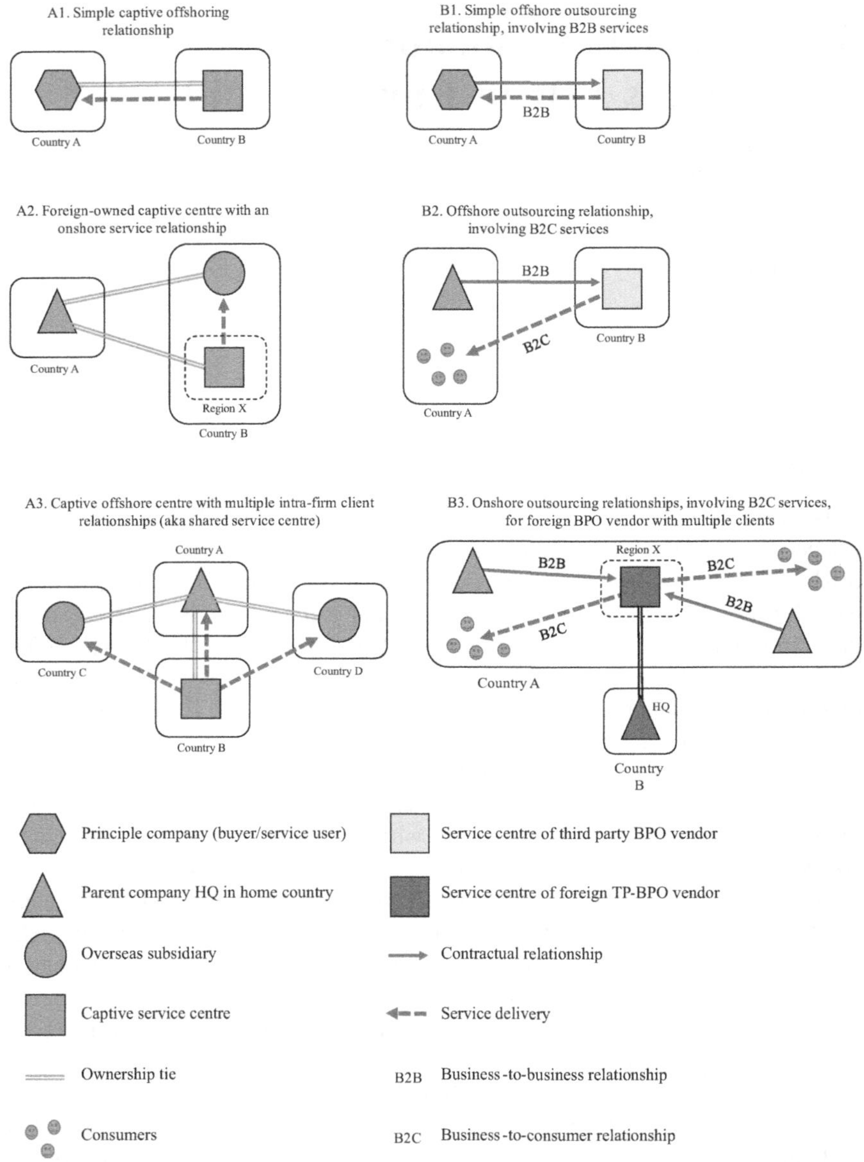

Figure 8.1 Some possible geographical and organisational configurations in offshore services

Source: Author.

the implications of this new 'round of investment' for host region development. Since many of these regions previously received significant inflows of manufacturing 'branch plant' inward investment – a phenomenon that spawned a large body of research from the 1970s to the 1990s (Firn, 1975; Hayter, 1982; Sonn and Lee, 2012; Watts, 1981) – there are also interesting historical comparisons to be made between these earlier rounds of manufacturing investment and the more recent wave of service-related FDI.

This chapter seeks to address the aforementioned issues, and in doing so to extend the geographical coverage of the service offshoring literature, via an exploratory case study investigation of offshore service-related FDI in Northern Ireland – a lagging old industrial region in the periphery of the UK and Western Europe. The study is concerned to understand the particular types of services projects attracted to this particular region and the ways in which these investments have (re)positioned the region within wider intra-firm and inter-firm networks and the international division of labour. It also seeks to consider some of the employment impacts of these projects on the host economy. Although, Northern Ireland has its own particularities – not least politically and socially – it is argued that this case study may have wider salience for regions with similar 'contextual dimensions', i.e. regions that have pursued, or plan to pursue, services FDI-led economic strategies in an attempt to address lagging economic performance.

The chapter is organised as follows. The next section reviews relevant prior literature on service offshoring and regional development and develops an analytical framework to guide the empirical analysis of the case. This is followed by an explanation of the study methodology and an overview of the case study region. The main empirical body of the chapter comprises three sections that: (1) Profile Northern Ireland's inward-investment performance since the mid-1990s and chart the rise of services projects; (2) Explore the 'structural characteristics' of the leading offshore services investors in the region (including various subsidiary attributes and their network positions); and (3) Examine some of the employment impacts of these projects on the region. The chapter concludes by considering the implications of the study and making some suggestions for further research.

Relevant literature and analytical themes

Offshoring and outsourcing, services FDI and host economies

Service offshoring and outsourcing can be understood in the context of various changes in business strategy and organisation, technology and international political economy. In the contemporary global economy, multinational firms are under intense competitive pressures that require them to seek cost efficiencies and improve their capabilities right across the value chain (Yeung and Coe, 2015). In pursuing these twin imperatives, business scholars observe an increasing tendency – facilitated by technological advances and the rise of emerging economies – for firms to 'fine-slice' their value-chains and 'unbundle' or 'functionally decouple' discrete activities and tasks from other related parts of the value chain and relocate them – both geographically, via 'offshoring', and organisationally, via 'outsourcing' (Buckley and Strange, 2015; Contractor *et al.*, 2010). Hence, in the case of various 'corporate services', there has been an upsurge in both captive offshoring FDI projects and also 'offshore outsourcing' to TP-BPO vendors, many of which have become sizeable multinationals in their own right (Gereffi and Fernandez-Stark, 2010; UNCTAD, 2004). From the point of view of host economies, this has presented a new opportunity to attract services FDI – in the form of captive centres and delivery centres operated by TP-BPO vendors.

A small but growing literature has sought to explore this phenomenon from the host economy's perspective. Most authors agree that the primary drivers of this phenomenon are cost-reduction – i.e. efficiency-seeking – and accessing appropriately skilled labour – i.e. resource-/talent-seeking. Therefore, the key location attractors for host economies are said to include pools of available and cost-efficient skilled labour, state-of-the-art ICT infrastructure, suitable office space at competitive rates, a stable and conducive business environment, and cultural affinity with target markets and investor home countries (Breathnach, 2000; Dossani and Kenney, 2007; Pandy and Rogerson, 2012; Sass and Fifekova, 2011). Some studies have highlighted the role played by local institutional actors in the process of strategic coupling with services production networks (Kleibert 2014; Hardy *et al.* 2011), a factor largely overlooked in the International Business literature (for example Doh *et al.,* 2008). Research evidence on the host economy impacts of offshore service-related FDI has produced mixed findings. Some authors have been broadly positive about the volume and quality of employment created in emerging locations (Benner, 2006; Kleibert, 2014; Sass and Fifekova, 2011) whereas others have been more sceptical, highlighting concerns about job quality and durability (Breathnach 2000; Hardy *et al.*, 2011). There is some agreement – at least in studies on Central and Eastern Europe – that the wider indirect employment, supply chain impacts and spill-over effects of such projects are likely to be quite limited, even by comparison to manufacturing branch plants (Hardy *et al.,* 2011; Micek *et al.,* 2011). Some authors have observed that service offshoring investments may exacerbate sub-national inequalities, due to their tendency to gravitate towards larger urban centres with abundant labour supply (Pandy and Rogerson 2012; Hardy *et al.*, 2011). Whilst the emergence of 'tier-2' offshoring destinations has also been documented in some countries (Kleibert, 2014), there is a suggestion that such locations might attract 'lower quality' operations (Kleibert, 2015).

Analytical framework development

Despite growing attention to service offshoring and outsourcing, there is no established, integrated framework for analysing this phenomenon from the host economy perspective. Therefore, to guide the analysis of the Northern Ireland case, a framework comprising 14 analytical themes relating to the study aims (grouped under three headings) was developed by synthesizing insights from the prior literature on FDI and regional development issues (for example Young *et al.,* 1994), a recent article by Kleibert (2015) and the subsidiary management literature (Paterson amd Brock, 2002) (Table 8.1, column 1).

The first six analytical themes describe various 'subsidiary attributes' (Table 8.1, heading I). The literature on FDI and regional development highlights the need to consider ownership issues, where a distinction can be made between domestic and foreign-owned operations and between multinational firms of different nationalities (Theme Ia), and the FDI entry mode, where a distinction can be made between greenfield and acquisition entries and between first round and repeat investments (Theme Ib) (Dicken, 2011; Sonn and Lee, 2012; Young *et al.*, 1994). The offshoring literature highlights the importance of the 'governance mode', either captive offshoring and offshore outsourcing (Sako, 2005; UNCTAD, 2004). From the host economy's perspective, a distinction should be made between captive centres and TP-BPO centres (Theme Ic). The type of services delivered by the subsidiary can be considered against the categories outlined by UNCTAD (2004) – i.e. call/contact centre services, share service centres providing back-office services and IT services – and Gereffi and Fernandez-Stark (2010) – i.e. information technology outsourcing, business process outsourcing and knowledge process outsourcing (Theme Id). An additional distinction can be made between subsidiaries that deliver services to private consumers and those that serve internal or business-to-business demand (Theme Ie). Finally, a subsidiary's scope can be described along various dimensions (Manolopoulos, 2008; Young *et al.*, 1994); since 'product scope' makes little sense in the context of service offshoring, the key dimensions to consider are the breadth of the task mandate (akin to functional value-added scope) and geographical market scope (Theme If).

The second set of four themes relates to the 'network position' of the subsidiary (Table 8.1, heading II). The first analytical theme is the 'locus of control', with a distinction being made between local and non-local control, and also between internal and external control (theme IIa). The 'branch plant economy' literature from the 1970s/1980s drew attention to the implicit risks to a regional economy arising from a high level of non-local control and dependence on distant decision-makers (Firn, 1975; Watts, 1981). The continued salience of this issue for service offshoring has been noted by Kleibert (2015), who also highlights another potentially important dimension. Due to the distinction between captive and outsourced governance models, there is an additional need to distinguish between internal (intra-firm) and external (inter-firm) forms of control. Captive centres are subject to non-local control by the parent firm's headquarters (as with a traditional branch plant) but delivery centres owned by foreign TP-BPO vendors are additionally subject to (local or non-local) external control, due to the inter-firm relationships between BPO vendors and their corporate clients. This latter form of control may be a particular concern for host regions due to the cost-driven nature of many buyer-vendor BPO relationships. The second theme under this heading, drawing on the subsidiary management stream (Manolopoulos, 2008), concerns the subsidiary's role and competences within its wider network context (Theme IIb). Young *et al.* (1994) suggest a subsidiary's host economy impact may depend on its

Table 8.1 Analytical comparison of two main subgroups among leading services foreign investors in Northern Ireland

Analytical categories	*Group 1: Captives with initial IT focus (n=4)*	*Group 2: TP-BPO vendors (n=6)*
I. Structural characteristics: subsidiary attributes		
a. Ownership (domestic/foreign nationality)	Foreign, exclusively US-owned	Foreign, mix of US, Indian and French-owned
b. Entry mode (greenfield/acquisition, single/repeat investments)	Mainly greenfield, followed by multiple repeat investments	Mix of greenfield and acquisition, with multiple repeat investments
c. Governance mode (captive/outsourced)	Captive centres	TP-BPO centres
d. Type of services, as per UNCTAD (2004) and Gereffi/Fernandez-Stark (2010) categories	Mainly IT services, some other shared services (UNCTAD); mainly IT offshoring, some knowledge and business process offshoring (G&FS)	(Inbound and outbound) call centres and back-office services (UNCTAD); Business process outsourcing (G&FS)
e. Nature of service delivery (internal/external, B2C/B2B)	To internal demand (parent firm and/or sister subsidiaries)	To private consumers on behalf of clients, sometimes directly to client firms (back-office tasks)
f. Breadth of task mandate and geographical scope	Initially narrow task mandate but subsequently broadened; expansion of geographical scope – e.g. from US parent to worldwide corporate network	Initially narrow task mandate with minimal broadening; mainly focused on onshore/nearshore provision for UK and Ireland market (but occasionally 'export' markets)
II. Structural characteristics: network position		
a. Locus of control (local/non-local, internal/external locations)	Internal, non-local control from parent HQs in developed world cities	Internal, non-local control from parent HQs in developed or emerging world cities *and* external, non-local control from client HQs in developed world cities
b. Distinctiveness of role (unique/ comparable) and competences (specific/generic)	Typically unique role and specific competences within parent firm (evolving over time)	Role comparable to other subsidiaries within parent firm and competences largely generic

Table 8.1 continued

Analytical categories	*Group 1: Captives with initial IT focus (n=4)*	*Group 2: TP-BPO vendors (n=6)*
c. Network embeddedness: strategic relatedness (high-low), connectivity (high-low), centrality (nodal-peripheral)	Not core node but high intra-firm connectivity; role supporting parent and sister subsidiaries suggests moderately high strategic relatedness	High strategic relatedness to parent firm; importance of host market may increase subsidiary status; lower strategic relatedness and more peripheral to client firms
d. Ego-network: location of peer subsidiaries	Often no direct peer in parent network; strong network connections to some world cities	Tier 1 and 2 cities in emerging economies and lagging regions in developed economies
III. Direct employment impacts		
a. Job quantity	Initially medium-sized projects with subsequent, sometimes significant, expansion; significant job numbers in context of regional IT labour market	Medium-sized or large projects, with sometimes rapid expansion; some very significant job numbers in particular local labour markets
b. Job quality (skill levels, graduate/non-graduate, wage levels)	Mainly managerial and skilled jobs; mainly but not exclusively graduate-level and above; mainly well paid jobs in the regional context	Mainly semi-skilled jobs; mainly non-graduate jobs with less emphasis on formal qualifications; mainly low paid jobs (with some pressure on terms and conditions)
c. Job stability/volatility and durability	Generally high stability and steady expansion; apparently high durability with no closures or significant down-sizing to date	Significant volatility in some companies and churn in the wider sector; some closures but projects more durable (less footloose) than might be expected
d. Sub-regional distribution	Mainly concentrated in Belfast (regional capital and largest city)	Spread between Belfast and some smaller cities and large towns (but not widely dispersed)

Note: The two main subgroups, as described in Table 8.1, were identified among the 12 leading services foreign investors (based on peak employment during 2000–13); the table provides a general summary and may conceal some variations between individual subsidiaries.

distinctiveness, with those having a unique role and specific competences being seen as more favourable than those with comparable 'sister' subsidiaries and generic competences. The concept of 'network embeddedness' from the global production networks literature is also relevant here (Henderson *et al.*, 2002). This suggests the importance of considering the subsidiary's position within the wider 'architecture' (structure and connectivity) of the (inter- and intra-firm) networks to which the subsidiary belongs (Theme IIc). Thus, Kleibert (2015) suggests 'network position' be gauged on a continuum from 'peripheral' to 'nodal'. Insights from the business literature suggest the 'strategic relatedness' between a subsidiary's competences and its parent company's goals might also be a relevant issue (Birkinshaw, 1996). A final concept that might be relevant to the discussion of a service subsidiary's network position is Mans' (2014) notion of an 'ego-network', which refers to the group of peer cities that occupy similar positions in the wider intra- and inter-firm corporate network. Thus, a subsidiary's ego-network may provide insights on the host region's role and position within the international division of labour (Theme IId).

The third part of the analytical framework concerns the direct employment effects of offshore service-related FDI projects (Table 8.1, heading III). Here, following the precedent of prior literature on FDI and regional development and more recent studies on service offshoring, the main sub-themes are: job quantity, job quality, job stability and longevity, and sub-regional equity issues (Young *et al.*, 1994; Dicken 2011; Sonn and Lee 2012; Kleibert 2015; Hardy *et al.*, 2011). For example, Sonn and Lee (2012) note that the 'branch plant' literature has been concerned with the quantity of jobs, the stability of employment and the quality of jobs (gauged by the ratio of managerial and skilled jobs vs. manual, semi/unskilled jobs). Young *et al.* (1994) suggest that job quality might be gauged by the percentage of graduate jobs, skill levels (high vs. semi-skilled) and wages (above or below sectoral or regional average). According to Gereffi and Fernandez-Stark (2010), wage levels may also provide the best available proxy for the level of value-added in a particular offshore service operation. Other potential employment-related effects on the host economy, including the upgrading of human capital and wage multiplier effects (Hardy *et al.*, 2011; Micek *et al.*, 2011) are acknowledged here but were beyond the scope of the present study.

Research design and data sources

Guided by the themes outlined above, the chapter employs a longitudinal case study approach to explore Northern Ireland's experience of offshore service-related FDI during the period from the mid-1990s to 2010 and beyond. The case study pays attention to the experience of the region as a whole but also, particularly, to 12 leading services foreign investors (embedded units of analysis), consistent with Markusen's (1994) notion of 'studying regions by studying firms'. These leading investors all undertook FDI into Northern Ireland and were engaged in the type of service activities and tasks identified by UNCTAD (2004) and Gereffi and Fernandez-Stark (2010). Note that domestic third-party outsourcing vendors operating in Northern Ireland (either locally or GB-owned) were explicitly excluded, as were captive operations operated by GB-owned companies and foreign investors that served the local Northern Ireland market only (e.g. retailers, banks, utilities). The case study was constructed from a number of different quantitative and qualitative secondary data sources, with triangulation being used (where possible) to corroborate facts and interpretations and improve reliability. The chief secondary data sources used in the project were:

1 A spreadsheet containing details of all government-assisted job-related investment projects in Northern Ireland over a 15-year period between the years 1995/96 and 2009/10 (obtained from Invest NI, the regional development agency). This was thoroughly analysed to identify key patterns and trends and the most significant foreign investors.
2 A series of employment data, extracted from the Equality Commission for Northern Ireland's Annual Monitoring Reports, which identified the number employees at every major services inward investor in the region between 2000 and 2013. These were used to chart the growth (and decline) of individual subsidiaries and overall employment trends.
3 An archive of regional and local newspaper articles, going back to the late 1990s, relating to each of the major services inward investors in the region, which was compiled using keyword searches of the LexisNexis database (following the precedent of economic geographers like Horner and Aoyama, 2009; Micek *et al.*, 2011). This provided a rich qualitative dataset, which complemented the aforementioned quantitative datasets. Detailed analysis of this dataset was conducted using themes identified from the extant literature (including those listed in Table 8.1). Key events in the development of each subsidiary were identified and explored (with the constraints of data availability).

Northern Ireland: attributes and prior inward-investment experience

Northern Ireland is a small lagging region of the United Kingdom (population approximately 1.8 million), situated in the north-west European periphery. Based on 2011 Census data, its capital and main city, Belfast, had an urban area population of 280,000 and the wider Belfast metro urban area had over 580,000 residents (NISRA, 2015), making it the second largest urban area (after Dublin) on the island of Ireland and broadly comparable in size to British cities like Sheffield, Bristol and Cardiff (ONS, 2012). The three next largest urban settlements (Derry/Londonderry, Bangor and Craigavon) each had 50–100,000 residents and a further 19 settlements had 10–30,000 residents but large parts of the region are rural (NISRA, 2015). The region has longstanding economic problems, by developed country standards, including relatively high unemployment, low economic activity rates, low wages, low productivity, an 'outdated' industrial structure and an over-dependence on the public sector (Barnett, 2009). The region had Objective 1 status (less than 75 per cent of the EU average GDP per capita) under the EU Structural Funding regime from 1988 until 2000, then 'transitional Objective 1' status from 2000–06. An additional factor are the 'Troubles' – a complex and deep-rooted 'ethno-political conflict' that lasted from late 1960s to late 1990s, which resulted in over 3,000 violent deaths and nearly 50,000 injuries (CAIN, n.d.). Since the late 1990s, however, there has been a high-profile 'peace process' resulting in a period of relative political and economic stability, although various legacies of the past remain.

Northern Ireland has significant prior experience with inward investment, which has been an important plank of industrial policy since the late 1950s (Teague, 1987). 'Branch plant' manufacturing investment in sectors such as textiles and clothing, man-made fibres, food processing, engineering and auto components dominated inward investment into the region during the 1960s and 1970s. However, the quality and sustainability of these investments attracted some criticism and there were waves of branch plant closures and significant job losses, notably in the early 1980s (Fothergill and Guy, 1990; Teague, 1987). In the late 1980s and early 1990s, there was some success in attracting new projects in more high-tech sectors like aerospace, electronics and telecommunications equipment but Northern Ireland's overall inward-investment performance during this period was inferior to other UK peripheral regions (Scotland, Wales, North-East England), and especially the Republic of Ireland (Crone, 1998). Since the beginning of the 'peace process' and the Belfast Agreement of 1998, local political and business elites have placed high expectations on inward-investment promotion as an engine for regional economic restructuring/modernisation and job creation. Although efforts to stimulate indigenous industry and entrepreneurship have received greater attention in this era, FDI promotion and investor support has remained a key priority for the regional development agency, Invest Northern Ireland (Invest NI, 2008).

Northern Ireland's inward-investment experience since the mid-1990s: a shift towards services

Data from the spreadsheet obtained from Invest NI, which covered all government-assisted inward-investment projects (including both FDI projects and investments from companies based in Great Britain) was used to profile Northern Ireland's experience over the 15-year period between 1995/96 and 2009/10. A key insight was that, beginning in the late 1990s, Northern Ireland's inward-investment profile switched notably towards services projects and gradually (though not totally) away from manufacturing projects. Figure 8.2 depicts the annual number of 'new jobs promoted' in assisted inward-investment projects by broad sector. New services jobs promoted increased notably from 1997/98 and exceeded the number of new jobs promoted in manufacturing in every year from 2000/01 onwards. Aside from a dip coinciding with the global recession around 2001/02–2002/03, the region typically attracted between 1,500 and 3,500 new services jobs promoted per year after 1997/98. Cumulatively, during the 15-year period to 2009/10, just under 30,000 new services jobs were promoted, in over 220 assisted projects, with the number of jobs and projects increasing in each of the three consecutive five-year sub-periods. To put these data in a wider context, Northern Ireland is said to have outperformed all other UK regions (though not

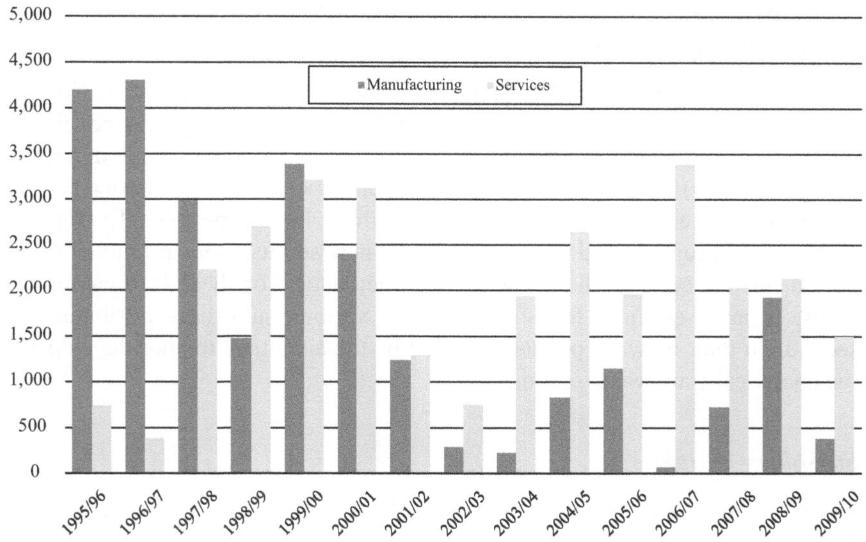

Figure 8.2 Number of new jobs promoted annually in Invest NI-supported inward-investment projects by broad sector, 1995/6–2009/10

Note: Data shown here relate to all assisted 'inward-investment' projects into Northern Ireland – i.e. including investments from the rest of the UK (GB) as well as FDI. Over this whole period, 76 per cent of all new jobs promoted (42,157) and 72 per cent of all assisted projects (n=389) were in foreign-owned (i.e. non-UK) companies. Also, note that these figures refer to the expected number of jobs associated with assisted projects, rather than the actual number of jobs created.

Source: Author's calculations from data supplied by Invest Northern Ireland.

the Irish Republic) in terms of jobs promoted in new FDI projects per capita between 2002/03 and 2007/08 (Barnett, 2009), and it also attracted more knowledge-intensive (including ICT) FDI projects and business services FDI projects in proportion to its gross value-added than any UK region outside south-east England during 2005–09 (fDi Markets, 2011). The Invest NI dataset reveals one further clue about the type of services projects and jobs being attracted to the region during this period. Invest NI categorised services projects into two broad segments, dubbed 'software and computer services' and 'business and financial services' (predominantly comprising various contact centre and back-office service operations). The former category accounted for 9,667 new jobs promoted in 121 projects over 15 years, whilst the latter accounted for 19,138 jobs in 89 projects (a residual category of 'other services' comprised a further 17 projects and 1,156 jobs promoted).

Structural characteristics of the leading offshore service-related foreign investors

More detailed insights on the nature of the services FDI attracted to Northern Ireland are provided by taking a closer look at the 12 leading services foreign investors (which were identified based on their peak employment during the period 2000–13). These 12 firms accounted for 22 per cent of all assisted services inward-investment projects received by Northern Ireland between 1995/96 and 2009/10 (47 projects) and 41 per cent of all new services jobs promoted in this period. In this section, the 'structural characteristics' of these leading services investors are examined focusing, first, on various subsidiary attributes and, second, on their network position (guided by the analytical themes developed in the analytical framework section).

Subsidiary attributes

Based on the various subsidiary attributes described in Table 8.1, it is possible to delineate four types of operation among the 12 leading foreign investors (Table 8.2). However, before looking more closely at these subgroups, it is worth making some overall observations. Collectively, these firms are involved in a broad range of service tasks from across the categories identified by UNCTAD (2004) and Gereffi and Fernandez-Stark (2010), including inbound and outbound customer contact, back-office administration, various IT support and software development tasks, and some knowledge-intensive shared services tasks. Both of the main governance modes identified in the literature are present, with five captive centres and seven outsourced BPO vendors. US firms are in a majority but the presence of French, Spanish, Japanese and Indian firms indicates a degree of diversity in the region's sources of services FDI. In terms of entry mode issues, note that seven of the firms first entered Northern Ireland as greenfield investors but five entered via acquisitions (of existing inward investors or local firms). An additional point of interest concerns the prevalence of repeat investment; each of the 12 leading foreign investors made multiple repeat investments in the region and they cumulatively accounted for over 40 assisted projects in the period under investigation (Figure 8.3). This shows that, far from being the end-game, an initial investment can often be the starting-point for an ongoing relationship between a foreign investor and the host economy.

Turning now to the four categories delimited in Table 8.2, the first sub-group comprises four US-owned subsidiaries that were initially (and in three cases primarily) involved in various IT, software or technology development and support activities (hereafter Group 1). Two of these subsidiaries, owned by insurance giants Allstate and Liberty Mutual, were initially established in the late 1990s; according to news media coverage from the time, the primary driver was the need to source appropriately skilled IT workers at competitive wage rates during a time of labour scarcity and rising costs in the United States (efficiency-seeking and talent-sourcing motives). These two cases, along with a later investment by NYSE

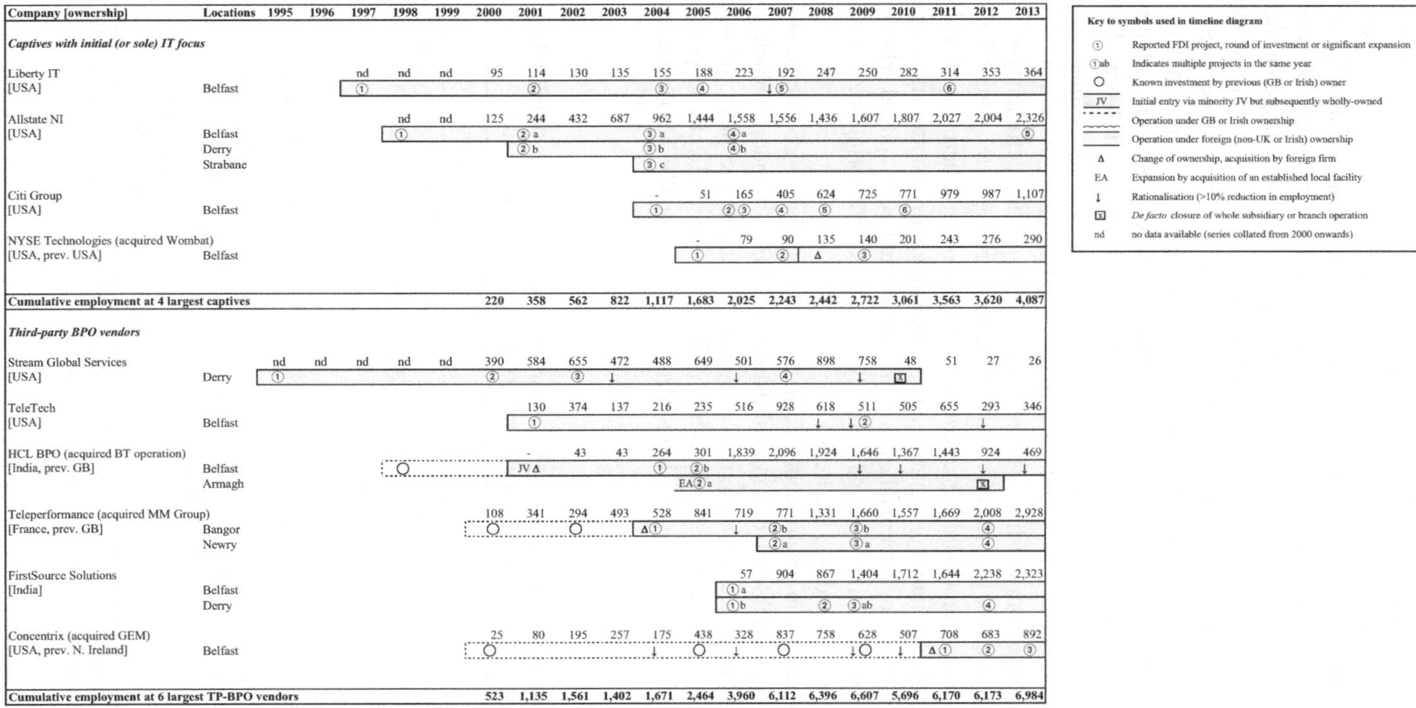

Figure 8.3 Timelines showing key events and employment data for four largest 'captive' and six largest TP-BPO foreign investors in Northern Ireland

Source: Author's compilation, based on various secondary data sources and Equality Commission Annual Monitoring Reports (employment data).

Technologies, fit within the classic definition of FDI and also the captive offshoring model (as depicted in Panel A1, Figure 8.1). Over time, the range of IT-related services undertaken by these subsidiaries has broadened, with some evolution into more sophisticated tasks, and some diversification into non-IT business processes. These subsidiaries have also widened their geographical scope from an initial focus on their US parent companies to supporting their parent firm and its subsidiaries across numerous countries. Thus, their organisational and geographical relationships have become more complex over time (as depicted in Panel A3, Figure 8.1). These subsidiaries are mainly involved in international trade in ICT-enabled services of the type described in the WTO GATS Mode 1 (but on an intra-firm basis). The case of Citigroup is slightly different. This subsidiary moved quickly beyond its initial IT development and support role, adding additional business process and knowledge process tasks, some of which were relocated to Belfast from the firm's London base (motivated by efficiency-seeking). The subsidiary now provides a variety of support services to the parent firm's corporate and investment banking operations in London and across the wider EMEA region, via a technology centre, a back-office operations centre, and a legal and compliance division. As a result, its organisational and geographical relationships are more complex than in a traditional dyadic captive offshoring scenario (as depicted in Panels A2 and A3, Figure 8.1).

The second major sub-group identified in Table 8.2 comprises six subsidiaries operated by US, French and Indian multinational TP-BPO vendors, which are mainly involved in delivering various (multi-channel) customer contact and relationship management activities, on behalf of their corporate clients, to private consumers within the UK and Irish markets, and occasionally across Europe

Table 8.2 Classification of 12 leading services foreign investors in Northern Ireland by main activity, governance mode, ownership and initial entry mode

	Captive centres	*Third-party vendors*
Investors initially involved in IT, software or technology development and support activities	*Group 1: US-owned captives with initial IT focus* Allstate NI (US) [GF] * Citi (US) [GF] * Liberty IT (US) [GF] NYSE Technologies (US) [ACQ]	*Type 3: IT outsourcing vendor* Fujitsu Services (Japan) [GF]
Investors primarily involved in contact centre (and other business process) activities	*Type 4: Captive centre with business process focus* Santander (Spain) [ACQ]	*Group 2: TP-BPO vendors* Stream Global Services (US) [GF] TeleTech (US) [GF] HCL Technologies (India) [ACQ] FirstSource Solutions (India) [ACQ] Concentrix Technologies (US) [ACQ]

Notes: Initial entry modes: [GF] = Greenfield, [ACQ] = Acquisition;.* indicates subsidiaries subsequently involved in various (non-IT) business process, shared services and knowledge process activities.

Source: Author's classification (typology developed from UNCTAD, 2004, and Sako, 2005).

(hereafter Group 2). These services include inbound contact centres (e.g. handling customer service enquiries, providing technical support), outbound contact centres (telemarketing/telesales) and some back-office administration tasks. According to newspaper content analysis, the corporate clients of these centres have spanned various vertical segments (for example banking, telecommunications, consumer electronics, utilities, retail and government) and included 'blue chip' companies like BT, HSBC, Deutsche Bank, Sony, EDF Energy, B&Q, Sainsbury, Dell Computer, Microsoft and Logitech. Although the relationship between these BPO vendors and their clients involves outsourcing, these cases do not match the description of offshore outsourcing seen in the existing literature and depicted in Panels B1 and B2 (Figure 8.1). Rather, the six foreign-owned TP-BPO vendors have undertaken FDI into the UK with market-seeking motives, seeking to establish an onshore/nearshore service delivery presence. This may reflect the demands of clients in the host market, in response to the widely reported customer backlash against distant offshore call centres. The decision to locate in Northern Ireland specifically then appears to reflect an efficiency-seeking motive, with the region offering a cost-competitive location within the UK context (in terms of wages, office costs and labour turnover rates). Thus, the subsidiaries in this category tend to have geographical and organisational relationships of the type depicted in Panel B3 (Figure 8.1); in this scenario, there is inter-regional but rarely international trade in services, yet the host region attracts inward FDI. An additional observation about the firms in this group concerns their initial entry mode. Whilst three of the companies came to Northern Ireland as *bone fide* new greenfield investors, it is notable that the others entered by acquiring an established contact centre from a domestic company (Figure 8.3). This trend can be viewed as part of a wider ongoing consolidation of the TP-BPO sector, both globally and within the UK market, and attempts by the acquirer firms to expand their 'global footprint'.

Finally, note that the third and fourth categories, which complete the typology in Table 8.2, include: a captive centre operated by the Spanish retail banking multinational Santander, which undertakes various back-office tasks and provides customer contact services to Santander's UK customers; and the local operations of the Japanese IT services multinational Fujitsu, which provide IT development and support services, on an outsourced basis, to corporate and government clients in the local and wider UK market.

Network position

Additional insights on the services FDI attracted to Northern Ireland, and the ways in which this has (re)positioned the region within wider international divisions of labour, come from considering the 'network position' of the two main subgroups against the themes outlined earlier. First, the 'locus of control', a key theme from the older literature on FDI and regional development, remains pertinent (Table 8.1, theme IIa). As was the case with previous rounds of 'branch plant' manufacturing investment, the region finds itself – via its service FDI inflows – subject to a significant degree of external control, with key decisions over significant numbers of jobs now resting in the hands of senior executives in distant global 'command-and-control' centres like New York, Boston, Chicago, Mumbai and Paris. In that sense, there may be concerns that region remains trapped in a 'dependent' situation. Further, in the case of firms in Group 2, there is also an additional layer of complexity to consider. These TP-BPO centres are run by multinational vendors, who compete for contracts from their client companies against rivals with broadly similar capabilities and largely (if not exclusively) on cost-efficiency. Such operations are therefore subject to an additional degree of 'external' control by their client organisations. This combination of intra-firm, non-local control and external control by client organisations, could be dubbed 'double dependence', and is likely to leave TP-BPO centres in a rather weak and dependent network position.

A second 'network position' issue concerns the distinctiveness of service subsidiaries, in terms of their role and competences, with their wider parent firm (Table 8.1, themes IIb). In the case of the TP-BPO centres (Group 2), it seems the types of services delivered are comparable to those delivered by other similar centres elsewhere within their parent company networks. It therefore seems unlikely that these centres possess any truly distinctive competences, although this claim awaits further investigation. This might leave these operations quite vulnerable to corporate rationalisations. However, during the period being studied, cultural, linguistic and time-zone compatibility with the UK and Irish markets, combined with favourable operating costs and demonstrated operational effectiveness – i.e. a favourable cost/capability ratio (Yeung and Coe 2015) – appears to have been sufficient for Northern Ireland to retain this type of corporate service activity. In contrast, the US-owned captives in Group 1 seem to fulfil much more distinctive roles within their parent firms, and have apparently – with parent company support – developed some valuable and specialised capabilities; for example, responsibility for developing and supporting software and systems for use across the corporate network (e.g. Liberty IT) or regional 'centres of excellence' for particular knowledge specialisms (e.g. Citi's legal and compliance unit).

The final point in this section concerns and the network embeddedness of subsidiaries, and builds on the 'ego-network' concept recently advanced by Mans (2014), which is held to offer insights into the status of particular places within the world economy (Table 8.1, themes IIc and IId). In the case of the US-owned captives in Group 1, it is not easy to identify equivalent 'peer cities' within their corporate networks, due to their apparently distinctive roles and specific competences. Rather, for these operations, the relevant ego-network seems to include important world cities such as London, New York, Boston and Chicago, since the Belfast-based subsidiaries in this group are connected to these headquarters cities by their provision of important support services. This moderate degree of network centrality and fairly high 'strategic relatedness' might be interpreted as an upward shift in the status of the region within the world economy. By contrast, in the case of TP-BPO centres (Group 2), the region finds itself part of an 'ego-network' that comprises cities in the lagging regions of developed countries (for example Cardiff or Middlesbrough in the UK; Halifax or St. John in the Atlantic provinces of Canada) and, more exotically, various cities in the emerging economies (for example Krakow, Buenos Aires, Cape Town, Cairo, Jaipur or Manila), since the multinational TP-BPO vendors have located similar service delivery centres in all of these places over the last decade. This latter ego-network may portray a less optimistic image of the region's (changing) role within the evolving international division of labour.

Direct employment impacts

The discussion in this final empirical section is organised around four themes: job quantity, job quality, job stability and longevity, and sub-regional equity issues (summarised in Table 8.1). Job creation remained a key regional policy imperative throughout the period under investigation. However, assessing the quantity of employment generated by services FDI is not straightforward, due to the lack of disaggregated official data on foreign firms. As reported earlier, around 30,000 new services 'jobs promoted' were reported by InvestNI over the period 1995/96–2009/10, with perhaps three-quarters of these being in foreign (rather than GB-owned) firms. However, 'jobs promoted' figures are notoriously unreliable and, in any case, not all of these jobs will have been in existence at the same time. For an alternative perspective, official Equality Commission data shows that the ten firms comprising Groups 1 and 2 in Table 8.2 employed a total of around 11,000 people by 2013, compared to only 750 in these same firms in 2000 (Figure 8.3). Although, these numbers may seem small by comparison with major service offshoring 'hotspots' in the emerging world, such locations are not a meaningful benchmark for a small peripheral region of 1.8 million people. The Kraków/Małopolska region of Poland, which is similar in size to the greater Belfast metro region, may be a more appropriate comparator. According to Micek *et al.* (2011), this Polish region – an alleged service offshoring 'hotspot' – had a total of around 9,700 jobs in captive and BPO centres by 2007. Another relevant yardstick, given the region's branch plant tradition is employment in foreign-owned manufacturing plants, which peaked at around 30,000 in 1979 (Fothergill and Guy, 1990) before declining to less than 20,000 by the early 2000s. When judged against these benchmarks, the number of service jobs attracted to Northern Ireland via FDI is certainly not trivial, and whilst the numbers may not be truly 'transformative' for an economy with nearly 700,000 private sector jobs, these projects can be said to have played a role in re-orienting the economy away from declining traditional sectors and partly offsetting the loss of manufacturing jobs. Also, taking a more fine-grained view, it is also worth noting that growth of IT-related employment at the four US-owned captives, from less than 250 jobs in 2000 to over 3,000 jobs by 2013 (Figure 8.3), represents a major contribution of new job opportunities to the regional IT labour market, since total employment in IT companies was only 4,000 in 1999. Finally, note that BPO centres have brought significant numbers of jobs to particular local labour markets across Northern Ireland, including some afflicted by high unemployment (e.g. Derry City, Newry).

Job quality in services FDI investors can be gauged by looking at wage levels, skill and qualification levels, and the balance between full-time and part-time jobs. The available evidence on these dimensions reveals some very significant differences in job quality between the two main types of investor (Group 1 captives versus Group 2 BPO centres). First, according to data reported by Invest NI (2008), only 29 per cent of the new jobs promoted in 'business and financial services' inward-investment projects (a good proxy for the BPO sector) between 2002/03 and 2007/08 paid above the regional private sector median wage, whereas the equivalent figure for 'software and computer services' projects was 86 per cent. This distinction is corroborated by evidence from news media coverage and adverts on regional job-finder websites. For example, several job expansion announcements quoted average gross annual salary figures that equate to two-to-three times the national minimum wage at Group 1 firms. By contrast, several of the BPO centres were found to be recruiting new entry-level employees at, or marginally above, the national minimum wage. Second, anecdotal evidence and media coverage of investment announcements suggests that the vast majority of jobs in the US-owned captive centres (Group 1) were skilled, graduate-level-and-above jobs. By contrast, the TP-BPO centres tended to require only basic secondary education, and in some cases they had relaxed the need for formal qualifications, placing more emphasis on 'soft skills' when recruiting new agents. Third, evidence from the Census of Employment 2013, suggests that 94 per cent of employees in 'computer programming, consultancy and related activities' firms were full-time, compared to only 83 per cent of employee jobs in 'activities of call centres' firms. Overall, the available evidence depicts a significant divergence in job quality between the two main subgroups, with the TP-BPO centres offering jobs of questionable quality and seemingly having much in common with the stereotypical manufacturing branch plant, whereas the US-owned captives seem to have brought mainly high quality jobs.

The job stability and longevity criterion also reveals some interesting contrasts between the two main subgroups. Among the four captives, there has been significant job stability and, in fact, steady (sometimes rapid) employment growth, seemingly linked to task mandate extension (Figure 8.3). The picture among the TP-BPO operations is considerably more mixed (Figure 8.3). Among the five investors that had been established in the region for more than three years, two firms have so far experienced employment growth and no notable rationalisations, but three exhibited notable volatility in employment levels, including several specific incidences of job shedding, two site closures (Stream in Derry and HCL's Armagh operation) (Figure 8.3). News media coverage supports the interpretation that BPO centres are inherently vulnerable to fluctuations in employment because the competitive project-based rivalry between BPO vendors is likely to result in the occasional contract losses.

A final issue is the sub-regional distribution of employment, which raises some spatial equity issues and policy challenges. A detailed spatial breakdown of employment in assisted services FDI projects is not available, so analysis of the 12 leading investors provides the best yardstick. A notable point is that all four

US-owned captives (Group 1) operate primarily in Belfast, the main urban centre, although the largest investor (Allstate) has also expanded to secondary sites in Derry City and Strabane (Figures 8.3 and 8.4). This preference for locating in Belfast may reflect the greater likelihood of being able to recruit sufficient numbers of skilled graduates and experienced managers in the largest urban centre, with its more diverse labour market, two universities and larger travel-to-work area. This issue presents an ongoing challenge to policy-makers who might wish to see a more equitable distribution of high quality jobs throughout the region. In slight contrast, among the six TP-BPO investors in Group 2, the two Indian firms (HCL and FirstSource) operated from a second location in addition to Belfast and two of the firms (Teleperformance and Stream) have never operated in Belfast (Figures 8.3 and 8.4). Thus, the geographical footprint of the TP-BPO centres is notably more dispersed than that of the captives (Group 1), extending to the second city (Derry) and some smaller urban centres (Bangor, Newry and Armagh) (Figure 8.4). This more dispersed pattern is thought to reflect the less stringent skill/qualification needs of these firms, their desire to secure access to additional workers away from the more-crowded Belfast labour market, and the possibility of securing lower labour and operating costs in the smaller urban centres. It is worth noting that the geographical dispersion of these BPO jobs is still much narrower than that associated with the manufacturing

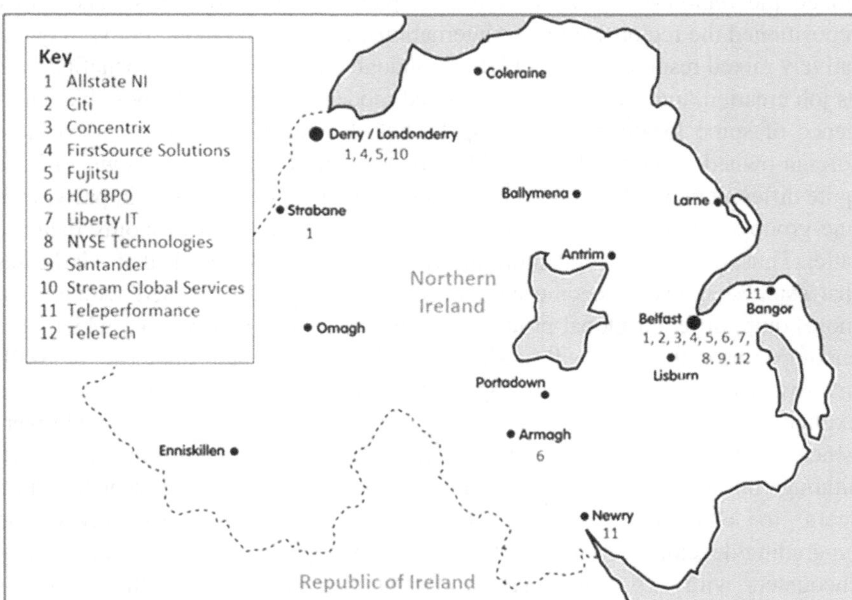

Figure 8.4 Map showing locations of 12 leading offshore services-related foreign investors

Source: Author.

branch plant era, when foreign investors were lured to a greater number of towns across the region. Thus, overall, despite some of their positive employment impacts on the regional economy, services foreign investors seem to have exacerbated existing sub-regional divisions of labour through their location preferences.

Conclusion

This chapter has extended the literature on the host economy implications of service offshoring and outsourcing and FDI into a new empirical context – lagging peripheral regions of developed economies – via a case study of the experience of Northern Ireland since the mid-1990s. The overall inward-investment experience of the region was examined and the structural characteristics and direct employment impacts of leading services foreign investors were explored. To support and structure this investigation, a new analytical framework was developed from the existing literatures on FDI and regional development, subsidiary management and service offshoring (Table 8.1). This framework may be transferable to other studies and should be trialled and adapted in different empirical contexts. The chapter has also elaborated on existing portrayals of service offshoring by identifying and delimiting some specific geographical and organisational configurations from the host economy's perspective (Figure 8.1).

In the Northern Ireland case, services FDI was found to have partially transitioned the region away from its branch plant manufacturing past, and partly repositioned the region within the international division of labour, but with qualitatively mixed results. Services FDI has undoubtedly brought some benefits, such as job creation, industrial restructuring and modernisation and the possible emergence of some locally-anchored capabilities. However, the two main types of foreign-owned service subsidiary identified in this region were shown to have quite different network positions and contrasting direct employment impacts, with one group seemingly having more intrinsic value to the host economy than the latter. Thus, one interesting feature of the Northern Ireland case is that it indicates that a small region can become simultaneously 'strategically coupled' with two (or more) quite different global production networks. This phenomenon, which here encompasses both 'low-road' (onshore TP-BPO centres) and 'higher-road' (skill-intensive captive centres) value-capture trajectories, does not seem to have been explicitly noted by any previous studies. It is suggested that this observed divergence may be related to the particular characteristics of the regional labour market, although this awaits further investigation. However, this rather mixed picture does seem to challenge simplistic binary distinctions between favourable (upgrading/development) and unfavourable (downgrading/ dependence) regional encounters with FDI and global production networks. Finally, the study has complemented existing studies of service offshoring destinations in the emerging world by revealing that lagging regions in the developed world can perform some distinct roles within evolving service production networks. For example, in Northern Ireland, foreign-owned TP-BPO centres perform an onshore/nearshore (rather than offshore) service delivery role, benefiting from geographical and cultural proximities to target (developed country) markets, whilst exploiting the cost advantages associated with lagging regions within these markets. Future research should dig deeper into the particular nature, basis and types of the 'strategic (re)coupling' between particular regions and different services global production networks (MacKinnon, 2012). There is also a need to explore and understand the possibilities for, and drivers of, subsidiary evolution in offshored and outsourced service centres. Combining insights from the subsidiary management stream within International Business with the global production network approach from economic geography would seem to offer a viable way forward.

Acknowledgements

This research underpinning this chapter was partly funded by the British Academy (Small Grant SG-51878). Valuable research assistance was provided in the early stages of the project by Iain Livingston, Brett Zych and Cherith Poots (students at Queen's University Belfast). The chapter is based on secondary data sources that are in the public domain. All interpretations/opinions and any errors are those of the author and not any organisation.

References

Barnett, R. (2009). *Independent Review of Economic Policy*. Department of Entreprise Trade and Investment (DETI) and Invest Northern Ireland.

Benner, C. (2006). "South Africa On-call": Information technology and labour market restructuring in South African call centres. *Regional Studies*, 40(9), pp. 1025–40.

Birkinshaw, J. (1996). How multinational subsidiary mandates are gained and lost. *Journal of International Business Studies*, 27(3), pp. 467–95.

Breathnach, P. (2000). Globalisation, information technology and the emergence of niche transnational cities: the growth of the call centre sector in Dublin. *Geoforum*, 31(4), pp. 477–85.

Buckley, P. J. and Strange, R. (2015). The governance of the global factory: Location and control of world economic activity. *Academy of Management Perspectives*, 29(2), pp. 237–49. doi:10.5465/amp.2013.0113.

CAIN. (n.d.). *Sutton Index of Deaths from the Conflict in Northern Ireland. Conflict Archive on the Internet, University of Ulster*. Retrieved from http://cain.ulst.ac.uk/sutton/index.html [accessed on 25 November 2015].

Contractor, F. J., Kumar, V., Kundu, S. K. and Pedersen, T. (2010). Reconceptualizing the firm in a world of outsourcing and offshoring: The organizational and geographical relocation of high-value company functions. *Journal of Management Studies*, 47(8), pp. 1417–33.

Crone, M. (1998). *Inward Investment and External Ownership in Northern Ireland Manufacturing: Historical Trends and Recent Developments*. Belfast: Unpublished paper, Northern Ireland Economic Research Centre.

Dicken, P. (2011). "Capturing value" within global production networks. In *Global Shift: Mapping the Changing Contours of the World Economy* (6th Edition., pp. 429–53). SAGE Publications.

Doh, J. P., Bunyaratavej, K. and Hahn, E. D. (2008). Separable but not equal: The location determinants of discrete services offshoring activities. *Journal of International Business Studies*, 40(6), pp. 926–43.

Dossani, R. and Kenney, M. (2007). The next wave of globalization: Relocating service provision to India. *World Development*, 35(5), pp. 772–91.

Egeraat, C. Van and Jacobson, D. (2004). The rise and demise of the Irish and Scottish computer hardware industry. *European Planning Studies*, 12(6), pp. 809–34.

FDI Markets. (2011). *Factsheet: UK regions' FDI Performance, 2005–09*. fDi Intelligence, Financial Times.

Firn, J. R. (1975). External control and regional development: the case of Scotland. *Environment and Planning A*, 7(4), pp. 393–414.

Fothergill, S. and Guy, N. (1990). *Branch Factory Closures in Northern Ireland*. Belfast: NIERC Report Series No 5, Northern Ireland Economic Research Centre.

Gereffi, G. and Fernandez-Stark, K. (2010). *The Offshore Services Global Value Chain*. Centre on Globalization, Governance and Competitiveness, Duke University. Retrieved from www.cggc.duke.edu/pdfs/CGGC-CORFO_The_Offshore_Services_Global_Value_Chain_March_1_2010.pdf [accessed on 26 November 2015].

Hardy, J., Micek, G. and Capik, P. Ł. (2011). Upgrading local economies in Central and Eastern Europe? The role of business service foreign direct investment in the knowledge economy. *European Planning Studies*, 19(9), pp. 37–41.

Hardy, J., Sass, M. and Fifekova, M. P. (2011). Impacts of horizontal and vertical foreign investment in business services: the experience of Hungary, Slovakia and the Czech Republic. *European Urban and Regional Studies*, 18(4), pp. 427–43.

Hayter, R. (1982). Truncation, the international firm and regional policy. *Area*, 14(4), pp. 277–82.

Henderson, J., Dicken, P., Hess, M., Coe, N. and Yeung, H. W.-C. (2002). Global production networks and the analysis of economic development. *Review of International Political Economy*, 9(3), 436–464.

Horner, R. and Aoyama, Y. (2009). Limits to FDI-driven growth in Ireland: a newspaper content analysis for investment, upgrading and divestment. *Irish Geography*, 42(2), pp. 185–205.

Invest NI. (2008). *Performance Information Report 2002/03-2006/07*. Belfast: Invest Northern Ireland.

Kleibert, J. M. (2014). Strategic coupling in "next wave cities": Local institutional actors and the offshore service sector in the Philippines. *Singapore Journal of Tropical Geography*, 35(2), pp. 245–60.

Kleibert, J. M. (2015). Global production networks, offshore services and the branch-plant syndrome. *Regional Studies*, pp. 1–15, doi:10.1080/00343404.2015.1034671.

MacKinnon, D. (2012). Beyond strategic coupling: reassessing the firm-region nexus in global production networks. *Journal of Economic Geography*, 12, 227–245.

Manolopoulos, D. (2008). A systematic review of the literature and theoretical analysis of subsidiary roles. *Journal of Transnational Management*, 13(1), pp. 23–57.

Mans, U. (2014). Revisiting city connectivity. *Journal of Economic Geography*, 14, pp. 155–77.

Markusen, A. (1994). Studying regions by studying firms. *The Professional Geographer*, 46(4), pp. 477–90.

Micek, G., Dzialek, J. and Górecki, J. (2011). The discourse and realities of offshore business services to Kraków. *European Planning Studies*, 19(9), pp. 1651–68.

NISRA. (2015). *2011 Census: KS101NI Usual Resident Population (Northern Ireland)*.

published 30 July 2015. Retrieved from www.nisra.gov.uk/ census/2011/results/key-statistics.html [accessed on 27 November 2015].

ONS. (2012). *2011 Census: KS101EW Usual Resident Population, Built-up Areas in England and Wales, Office for National Statistics*. Retrieved from www.ons.gov.uk/ons/rel/census/2011-census/key-statistics-for-local-authorities-in-england-and-wales/index.html [accessed on 27 November 2015].

Pandy, W. R. and Rogerson, C. M. (2012). The economic geography of South Africa's call centre industry. *Urban Forum*, 23(1), pp. 23–42.

Paterson, S.. and Brock, D. (2002). The development of subsidiary-management research: review and theoretical analysis. *International Business Review*, 11(2), pp. 139–63.

Sako, M. (2005). *Outsourcing and Offshoring: Key Trends and Issues*. Background Paper prepared for the Emerging Markets Forum, Oxford Said Business School, November.

Sass, M. and Fifekova, M. (2011). Offshoring and outsourcing business services to Central and Eastern Europe: Some empirical and conceptual considerations. *European Planning Studies*, 19(9), pp. 1593–1609.

Sonn, J. W. and Lee, D. (2012). Revisiting the branch plant syndrome: Review of literature on foreign direct investment and regional development in Western advanced economies. *International Journal of Urban Sciences*, 16(3), pp. 243–59.

Teague, P. (1987). Multinational companies in the Northern Ireland economy: an outmoded model of industrial development? In P. Teague (ed.), *Beyond the Rhetoric: Politics, the Economy and Social Policy in Northern Ireland* (pp. 160–182). London: Lawrence and Wishart.

UNCTAD. (2004). *World Investment Report 2004: The Shift Towards Services*. New York and Geneva: United Nations Conference on Trade and Development.

Watts, H. D. (1981). *The Branch Plant Economy: A Study of External Control*. Longman.

Yeung, H. W. and Coe, N. M. (2015). Toward a dynamic theory of global production networks. *Economic Geography*, 91(1), pp. 29–58.

Young, S., Hood, N. and Peters, E. (1994). Multinational enterprises and regional economic development. *Regional Studies*, 28(7), pp. 657–677.

9 New kids on the block?

The emerging global mosaic of cultural production

Robert C. Kloosterman and Rosa Koetsenruijter

> Culture follows wealth, prestige and glamour. As the US share of these declines, the world's viewers may come to prefer Sleepless in Seoul to Sleepless in Seattle.
>
> Christopher Caldwell (2012), 'Gangnam stylishly debunks US myth' *Financial Times*

Introduction

For those accustomed to the dominance of Western pop music, the autumn of 2012 contained quite a surprise. On November 24th of that year, the video of 'Gangnam Style' by the South-Korean artist Psy became the most watched video on YouTube when it surpassed the music video for 'Baby' by Justin Bieber from Canada. At the end of that year, 'Gangnam Style' was a number one hit in more than 30 countries, including Australia, Canada, France, Germany, Italy, Russia, Spain, and the United Kingdom (https://en.wikipedia.org/wiki/Gangnam_Style). The question whether 'Gangnam Style' was just a novelty song by a mere one-hit wonder or something bigger, transcends the importance of the individual artist Psy by far. Are we, then, witnessing the emergence of a new global geography of cultural industries which contests the hegemony of Western countries in the same way as South (East) Asian countries have challenged them in the field of manufacturing and now increasingly in producer services as well?

This question, explicitly triggered by 'Gangnam Style', was very eloquently posed by Christopher Caldwell (2012) in the *Financial Times*. Others have also looked at the role and the potential of cultural industries for economic development. The UNCTAD (2008, 2012, 2013) published a series of important reports on the creative economy covering both developed and developing economies stressing the opportunities offered by cultural industries. In these reports, various segments of the cultural industries are analysed especially with regard to their opportunities for growth and export and a number of case studies provide more in-depth views.

Sociologists, geographers, economists and other academics have also offered case studies on a number of aspects of the changing global division of labour regarding cultural industries. There are more abstract theoretical analyses (e.g. Appadurai, 1990), empirical studies on countries (e.g. Keane *et al.*, 2005, on

Asian and Latin American countries; Yusuf and Nabeshima, 2005 on creative industries in East Asia; Keane, 2004, on China; Shim, 2006, on South Korea; Straubhaar, 2010, on China and India), case studies on specific cultural industries (e.g. Bautès and Valette, 2004, on painting in Rajasthan; Power and Hallencreutz, 2002; Brandellero and Pfeffer, 2011; Watson, 2012, on music production; Krätke, 2003, and Hoyler and Watson, 2012, and Thussu, 2013, on media), and studies on cities (solely non-Western or broader) as centres of cultural industries (O'Connor, 2004, on St. Petersburg; Pratt, 2012, on global cities and cultural industries; Ren and Sun, 2012, and Yingying *et al.*, 2012 on Beijing; Kong *et al.*, 2015 on East Asian cities).

In this chapter we offer a preliminary empirical exploration of the role and importance of cultural industries in selected developing countries which evidently can be labelled as emerging economies, namely the East Asian Tigers and the BRICS countries. We, thus, depart from a relationship between the stage of economic development and the size and economic significance of cultural industries based on the assumption that higher levels of income and higher levels of purchasing power generate a higher demand for cultural products. We, therefore, limit ourselves to those countries which have already clearly moved up the ladder and compare their position with that of the established leaders of the global cultural production system, the United States and the United Kingdom. We depart from the definition used by the UNESCO (UNCTAD, 2008, 2012, 2013) which describes cultural industries as those activities 'which combine the creation, production, and distribution of goods and services that are cultural in nature and usually protected by intellectual property rights'. This definition has been widely adopted (Pratt, 2013). More specifically, we focus on four key cultural industries: film, music production, architectural design and fashion, and assess their importance from a comparative perspective. We explore the significance of these cultural industries on three levels of aggregation – the national, the urban, and the sectoral level – as each of these offer a distinct window on the significance of cultural industries.

Getting data on cultural industries is often already quite hard for many developed countries, this problem becomes much worse when one includes developing countries and even more so, if one wants to compare them across borders. By looking at three levels of aggregation and by using different data sources (employment, export, turnover, reputation), we are, however, able to identify broader trends regarding the global division of labour in cultural industries. Our aim, hence, is to contribute to the debate on the (potential) role of cultural industries in emerging economies as drivers of development by providing an initial empirical investigation of the global geography of cultural industries and by being able to identify main trends.

We provide a very brief summary of strategic analytical issues pertaining to cultural industries (Section 2). We then continue by explaining our methodology (Section 3). The next three sections cover the evolution of cultural industries on the national level (Section 4), the urban level (Section 5), and the export (Section 6). We conclude by discussing the implications of our findings (Section 7).

Cultural industries as drivers of economic development

Culture is a defining characteristic of human existence and every human society produces, in one way or another, culture – be it music, paintings, clothing, furniture or other goods and services strongly imbued with symbolic meanings. Cultural industries, however, are more specific as they produce such symbolic items under capitalist conditions. Culture, then, becomes commodified and, therefore, produced for markets to be sold for profit. In this sense, cultural industries can be viewed as any other economic activity aimed at markets. In many developed economies, cultural industries have become ever more important as drivers of economic growth and employment (UNCTAD, 2013). Increasing incomes in combination with shifts in consumer preferences towards more distinctive products on the demand side interacting with the rise of flexible-specialisation production technologies which are able to make small series on the supply side, have been crucial factors in the structural expansion of cultural industries (Pratt,1997, 2013; Scott, 2000; Kloosterman, 2004, 2015).

In his pioneering book *The Cultural Economy of Cities*, Allen Scott (2000) has outlined the general spatio-organisational structure of contemporary cultural industries. Due to rapidly changing and fragmented markets, short product life cycles of products, risks of product failure are relatively high. Economies of scale are, consequently, often hard to achieve and economies of scope instead are of strategic importance as they allow for swift responses to changes in demand. This vertical disintegration of the value chain thereby creates opportunities for small, specialised firms. These small firms are intricately connected to each other through myriad input and output linkages or so-called traded interdependencies (Storper, 1997). Notwithstanding the predominance of small and medium-sized firms, we do find (even very) large firms in cultural industries – for example the Walt Disney Company or Bertelsmann – which are able to use their size to fund large expensive projects (such as blockbuster films) and exploit economies of scale in marketing and distribution. However, these large firms often function as systems houses (Scott, 2000) which outsource much of their creative work to networks of small firms.

The organisational characteristics of cultural industries are recursively related to their spatial articulation. To reduce the relatively high transaction costs associated with these kind of linkages, firms tend to concentrate spatially in clusters of related activities (Scott, 2000). Clustering not only reduces transaction costs between firms, but also helps to minimise search costs for both firms and workers through the formation of a local specialised labour pool. In addition, clustering also generates other types of agglomeration economies or untraded interdependencies by enabling the spill over of knowledge which fosters innovation as well the creation of an infrastructure of dedicated institutions (for example, with respect to education, quality control, and marketing, see Scott, 2000, 2004). Cultural industries are, hence, in most cases clustered in urban settings and, moreover, especially in large cities, and, to descend one more spatial level, in particular neighbourhoods. This model of cultural industries has been widely documented in recent years (Hall, 1998; Scott, 2000, 2004, 2008, 2010; Currid, 2007; Kloosterman, 2008, 2010; Currid and Williams, 2010; Pratt, 2012; UNCTAD 2013). Here we use this model as an analytical point of departure although we are very much aware that cultural industries vary widely as they cover '…many activities from film, to radio, television and new media' and 'they all have different regulatory and organizational dynamics, as well as unique markets' Pratt (2012: 268–9).

Given the preponderance of small firms in cultural industries, the initial financial outlay to start a business there may be relatively low. In addition, the typical main input of cultural industries is not so much high-tech as high-concept knowledge (frequently tapping into local traditions), often, but not necessarily in combination with high craft skills. Accordingly, we expect that, in principle, entry barriers for firms in cultural industries are relatively low when compared to economic activities as manufacturing or producer services. On the basis of this, we can also expect that cultural industries offer opportunities for developing countries so they can build on their particular kinds of cultural capital and use their craft skills while only a relatively small amount of start-up capital is required. However, while entry barriers may be low, access to global markets might be more difficult for cultural industries in developing countries as distributions and marketing on an international scale often requires large firms or systems houses with ditto resources. The established systems houses may even actively bar newcomers.

Below, we will explore the significance of cultural industries in developing countries. Nowadays, cultural industries are seen as not just 'secondary' or 'dependent' on other economic activities but serve as autonomous drivers of economic development catering to domestic and export markets (cf. Pratt, 2012). According to the UNCTAD's (2013: 10) *Creative Economy Report* 2013, developing countries are already benefiting from the growth of cultural industries and 'when the creative sector becomes part of an overall development and growth strategy, it can contribute to the revitalization of the national economy where hybrid and dynamic economic and cultural exchanges occur and innovation is nurtured.' Scott (2000: 209), when focusing on the urban level, stated that '[a]s things now stand, New York and Los Angeles appear to be well in the forefront of commercialized cultural production for world markets, especially in regard to the key sectors of film, television, and popular music'. He anticipated, however, 'a global but polycentric and multi/faceted system of cultural production straddled by large multinational corporations – to make its appearance over the course of the twenty-first century' (Scott, 2000: 211). Can we already observe the contours of this global system of cultural production when looking at emerging economies which have already attained relatively high levels of GDP (per capita) and of economic development more in general? Or is the relationship between stages of economic development and prominence in cultural industries more complex?

Methodology and data

Our aim here is to assess the role of cultural industries in emerging economies and compare this with the global leaders in cultural production, the United States and the United Kingdom, which have dominated the global cultural production system in the second half of the 20th century (UNCTAD, 2008). We have selected two groups of emerging economies. The first group is that of East Asian Tiger economies – Hong Kong (which became a so-called *special administrative region* of the People's Republic of China in 1997), Singapore, Taiwan and South Korea – which experienced rapid industrialisation and strong economic growth from the mid-1950s (Hong Kong) and the 1960s (the others) onwards (Chang, 2002; Frieden, 2006; Dixon, 2009). The other group of manifestly emerging economies are the BRICS. These five countries – respectively Brazil, Russia, India, China, and South Africa – have been more recently marked as countries with both a strong economic emergence and large populations and, hence, potential important players in the global economy (O'Neill, 2011). These eight emerging economies (we consider Hong Kong as a city within China, and not as a separate national entity) will be compared the named established global cultural powers

We use the UNESCO definition of cultural industries as a general point of departure. However, when constructing datasets on the national and the urban level we have to rely on the definition of cultural industries used by the national and local statistical agencies. These definitions may differ per country and even per urban region although they tend to cover a similar set of cultural industries (UNESCO, 2008). For this reason, comparing them can only be indicative.

To explore the position of cultural industries in the selected emerging economies and the two leaders from a comparative perspective, we focus on four key cultural industries: film, music, architecture, and fashion. Ideally, we would like to have time series on employment, number and size of firms, turnover, export, added-value broken down by subsectors on a national level as well as on the level of individual cities. These data are already often difficult to obtain for most developed economies, let alone for developing countries and this holds even more true on the city level. Systematic international comparisons based on uniformly defined data with respect to cultural industries are, therefore, in most cases not possible. We, consequently, had to opt for an eclectic approach and have compiled data from various sources to construct a multidimensional picture of the role of cultural industries in emerging economies.

On the national scale, most data was collected from government websites and cultural reports, where possible data pertaining to the urban scale was gathered from these sources as well. When researching each of the four cultural industries, we aimed to find listings from official sources – from national statistical agencies to data from industry associations such as *BD online* on the ten largest architecture firms or from the *Recording Industry Association* of Japan on the global music sector. Gathering data on individual cultural industries within cities proved even more difficult. For this, we used *The World Cities Culture Report* (Mayor of London, 2014) as a starting point. This *Report* featured emerging as well as developed cities and provided information on the cultural sector comprising both the production and the consumption of culture.

The national level

Whichever listing of the best-selling films or music artists one looks at, the United States and the United Kingdom almost always come out on top as the countries of origin. This dominance of the United States can be attributed to a combination of general economic strength, the 'commercial know-how and energetic worldwide marketing strategies of American corporations' (Scott, 2000: 210), having a world language as the native tongue, and of soft power: 'the ability to shape the preferences of others through appeal and attraction' (Nye, 1990). The strong position of the United Kingdom can be partly related to being part of the same linguistic (and, up to some extent, cultural) realm as the United States (Pratt, 1997). Its global leadership is also based on its deeply rooted traditions of literature from Shakespeare onwards to Harry Potter and to its more recent ability to absorb American music culture while giving it its own particular twist (from the Beatles to Adele) thereby creating a more general cultural buzz ('Cool Britannia') comprising not just music and other audiovisual sectors (film, TV) but fashion as well (Sandbrook, 2015). Success breeds success and when cultural industries become successful, a path-dependent process may be set in motion attracting ever more talent, capital, and customers (Scott, 2000).

Based purely on revenues, the United States has indeed by far the largest cultural sector, whereas when one looks at the percentage of GDP generated by the cultural industries, the United Kingdom outranks all other countries. In Figure 9.1, the revenues and the share of GDP are shown for the top dogs United Kingdom and the United States as well as for Brazil, Taiwan, and Singapore, the only developing countries of our sample for which both sets of data were available. As we can see, the shares of cultural industries of the GDP of the three emerging economies are already substantial and, relatively speaking, in the same league as the United States. The United Kingdom is clearly an outlier with its very high share. The United States, on the other hand, are very much an outlier when looking at the absolute number of the revenues.

Cultural industries, aside from contributing to the GDP, generate employment as well. Table 9.1, below, depicts the total number of workers within the cultural industries, and the percentage of national employment in the cultural industries. Here, cultural industries not solely refer to the four key industries discussed in more detail below, but to the whole cultural sector comprising the production of cultural goods and services such as arts and crafts, architecture and design, audiovisual production, publishing and fashion design. As can be seen, the United States have (again) the largest cultural sector, now in terms of the numbers of workers employed within the cultural industries. When considering the share of the cultural sector in the total national employment, the United Kingdom is once again an outlier.

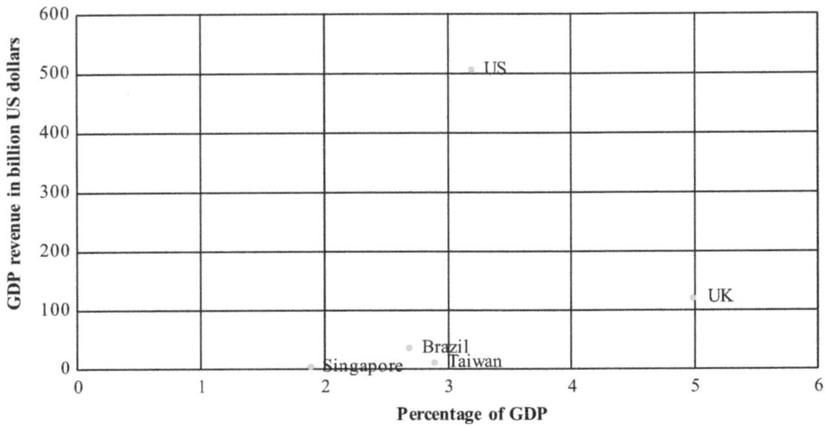

Figure 9.1 GDP revenue and percentage of GDP generated by the cultural sector of selected developed and emerging economies (Brazil: 2011; Singapore: 2000; Taiwan: 2000; United States: 2011; United Kingdom: 2013)

Sources: National Endowment for the Arts, 2013; DCMS, 2015; Cultural and Creative Industry Promotion Team, 2004; Sistema Firjan, 2012; Cho and Hoo, 2003; Kwon and Kim, (2013).

Table 9.1 The percentage of workers employed in the cultural sector and the absolute number of employees within the cultural industries for the selected developed and emerging economies (Brazil: 2011; Singapore: 2000; Taiwan: 2000; United States: 2011; United Kingdom: 2013)

Country	*Percentage of total national employment*	*Number of workers (× 1,000)*
United States	N/A	2,000
United Kingdom	5.6	1,700
Taiwan	3.5	325.5
Singapore	2.2	47
Brazil	1.7	810

Sources: National Endowment for the Arts, 2013; DCMS, 2015; Cultural and Creative Industry Promotion Team, 2004; Sistema Firjan, 2012; Cho and Hoo, 2003; Kwon and Kim, 2013.

Film, music, architecture and fashion; how do the emerging economies rank?

The significance of cultural industries in the selected emerging economies is, it seems, already quite substantial albeit not as big as those of the two global cultural leaders. We will now focus on the four key cultural industries to assess if this is holds on the level of individual subsectors.

Film

Those who assume that the United States produce the most films yearly, are mistaken. According to the UNESCO Institute for Statistics (UIS) (2012), India was the largest film-producing country in 2011 based on the number of films.[1] Table 9.2, shows in addition that besides India, two more emerging economies (China and South Korea) are listed in the top-ten feature film-producing countries.

Furthermore, when looking at the revenues of best-selling films, the United States do come out on top. Even though the United States do not produce the most films yearly, its revenues of the best-selling film (*Avatar*) are much higher than Bollywood's best-selling film (*PK*). *Avatar* has thus so far generated more than 2 billion dollars worldwide, while *PK* has generated a mere 120 million dollars (Boxofficemojo, n.d.; Boxofficeindia, 2015).

The fact that the revenues of the Indian films are substantially lower than those of the US films could be due to the fact that American films are featured worldwide (including on the lucrative markets of developed countries), whereas Bollywood mainly caters to the domestic Indian market (and the Indian diaspora) as well as the regional South Asian market (Straubhaar, 2010).

Table 9.2 Top ten film-producing countries in 2011

Country	*Number of films*
1. India	1,255
2. The United States	891
3. China	584
4. Japan	441
5. The United Kingdom	299
6. France	272
7. South Korea	216
8. Germany	212
9. Spain	199
10. Italy	155

Source: UIS, 2013.

Music

The music industry comes in many shapes and forms; live, recorded, digital and streamed. Over the years, the industry has first seen a decline in global recorded sales and thereafter a decrease in digital music, according to the *Financial Times* (Hunter Tilney, 2015, Garrahan, 2015). Notwithstanding problems with intellectual property rights due to piracy, a substantial amount of money is still made in churning out CDs and even vinyl. The global recorded music sales (physical and digital combined) generated over 15 billion dollars in 2014 (RIAJ, 2015).

When we look at Table 9.3, it can be seen that, first of all, developed economies are still global leaders both with respect to the production of music in the digital and the physical format. Brazilian and South-Korean artists have hitherto generated a market share of 2 per cent in the digital (South Korea) and physical (Brazil) global music market. The already significant market share of Brazil within the global music market can be traced back to the 1970s when the government launched an import-substitution programme (Straubhaar, 2010: 257). The Brazilian market share is as well in accordance with Straubhaar's conclusion stating that Brazil is 'increasingly autonomous in media production' (Straubhaar, 2010: 253).

Table 9.3 The physical and digital music industry in percentages of the total global market in 2013

Physical market shares in 2013		*Digital market shares in 2013*	
1. Japan	31	1. United States	45
2. United States	17	2. United Kingdom	10
3. Germany	13	3. Japan	8
4. France	8	4. Germany	5
5. United Kingdom	7	5. Australia	4
6. Canada	7	6. France	4
7. Australia	2	7. Canada	4
8. Italy	2	8. Sweden	2
9. Brazil	2	9. South Korea	2
10. Netherlands	1	10. Norway	1

Source: RIAJ, 2015.

Architecture

Films and music are seen as cultural industries that tend to be bounded by cultural proximity and consequently may encounter difficulties when transcending its particular cultural setting (Straubhaar, 2010). The high-end of architectural design, on the other hand, seems to be able to cross cultural boundaries more easily given its role as a marketing tool to brand (global) cities (McNeill, 2009; Ren, 2011: 4–5). Several design and architecture websites, and blogs publish rankings of the best architects (Gibson, 2012; Henderson, 2012 and Ford, n.d.). These listings are dominated by Western architects (cf. Ren, 2011). The architects from elsewhere, such as Ieoh Ming Pei (China), Cesar Pelli (Argentina) and the late Zaha Hadid (Iraq) and Oscar Niemeyer (Brazil), however, typically have been educated at institutions in Western countries and have subsequently set up their practices there as well.

Besides individual prominent architects, leading firms within the industry are assessed here as well. *BD Online* yearly releases a report stating the top-100 of the largest architecture firms worldwide, inter alia based on the number of architects employed. The top-ten largest architecture firms in 2013 are shown below in Table 9.4.

Table 9.4 Top ten largest architecture firms in 2013, based on the number of architects employed

Firm	*Number of architects*	*Country of origin*
1. Aecom	1,370	United States
2. Gensler	1,346	United States
3. IBI Group	1,129	Canada
4. Nikken Sekkei	1,109	Japan
5. Aedas	1,078	China
6. Perkins & Will	771	United States
7. DP Architects	726	Singapore
8. HOK	715	United States
9. Samoo Architects & Engineers	712	South Korea
10. Foster & Partners	646	United Kingdom

Source: WA100, 2014.

Wallis Simons (2014) remarked that Asian architecture firms are on the rise. In 2014, a Vietnamese firm, *a21 Studio*, won the *Building of the Year Award*, at the yearly world architectural festival in Singapore. At that same festival the Asian submissions increase yearly (Wallis Simons, 2014). This trend of upcoming Asian architecture firms can as well be seen in Table 9.4, in which three firms from emerging economies are listed, all originating from Asia.

Fashion

Several internationally renowned fashion designers are by different blogs and websites listed as the world's best fashion designers (ByKoket, 2013; Erkan, 2013; International Business Times, 2011; CEO World Magazine, 2011). In these listings, famous haute couture designers like Karl Lagerfeld, Giorgio Armani, Donatella Versace, Ralph Lauren, Tom Ford, Stella McCartney, and Raf Simons are always featured. Almost all the designers featured in these listings originate from Western Europe or the United States. Other listings (ByKoket, 2013 and CEO World Magazine, 2011) featured designers from Brazil, Colombia, Thailand, Greece and Russia as well. Just as the prominent architects, these designers have mostly moved to the United Kingdom, France or the United States.

Because 'world's best fashion designers' is very much a matter of taste, we also present the largest fashion and apparel brands and corporations (Table 9.5 and Table 9.6). Looking at the top 15 brands based on revenue, no emerging economies are listed, and the same can be said about the top ten leading apparel brands. In these markets, marketing and distribution are key and these activities are very much prone to economies of scale thus favouring large firms with extensive global networks. This creates considerable barriers of entry for newcomers and helps to strengthen the position of the established firms.

Table 9.5 Top 15 fashion brands in 2013, based on annual revenue in US dollars

Brand/corporation	*Annual revenue dollars (billion)*	*Country*
1. Louis Vuitton Moët Hennessy	37.1	France
2. Hennis and Mauritz	18.8	Sweden
3. Kering	15.65	France
4. The Gap	15.65	United States
5. Christian Dior	11.9	France
6. Richemont	11.8	Switzerland
7. Estée Lauder	9.7	United States
8. Phillips-Van Heusen Corp.	6.04	United States
9. Coach Inc.	4.76	United States
10. Levi Strauss	4.67	United States
11. Abercrombie & Fitch	4.51	United States
12. Jones Group	3.8	United States
13. Tiffany and Co.	3,79	United States
14. Hermès	3.68	France
15. American Eagle Outfitters	3.48	United States

Source: www.statisticbrain.com/fashion-industry-statistics

Table 9.6 Top ten leading apparel brands in 2015 based on brand value in US dollars

Brand	*Brand value in dollars (billion)*	*Country*
1. Nike	29.7	United States
2. Zara	22.0	Spain
3. Hennes and Mauritz	13.8	Sweden
4. Uniqlo	8.07	Japan
5. Next	5.97	United Kingdom
6. Ralph Lauren	5.6	United States
7. Adidas	4.6	Germany
8. Hugo Boss	4.3	Germany
9. Lululemon	2.9	Canada
10. Tommy Hilfiger	2.6	United States

Source: www.statista.com/statistics/267931/brand-value-of-the-leading-10-apparel-brands-worldwide

The United States and the United Kingdom are still clearly global leaders in cultural industries when one looks at the contribution to the GDP and the absolute and relative size of employment. Exploring the four key cultural industries corroborated this finding. What we do see, however, is that in music Brazil and South Korea are already among the top-ten producers, whereas the film industry in India is evidently an exceptional case producing the highest number of feature films and an obvious world player. According to Straubhaar (2010, 258): 'In the 1990s and 2000s, Indian film in particular began to reach much more global markets, following its diaspora, but going well beyond it, to draw in viewers from the larger audiences of many countries, including the USA.'

The urban level

Cultural industries tend to concentrate in large cities due to the salience of agglomeration economies (Scott, 2000, 2004; Currid, 2007; Currid and Williams, 2010 Pratt, 2012). Therefore, the capital and/or prominent cities within the emerging economies have been selected for the exploration of the urban cultural sector; São Paulo, Rio de Janeiro, Moscow, Mumbai, Beijing, Hong Kong, Shanghai, Johannesburg, Taipei, Singapore and Seoul. This selection is as well based on the cities incorporated in *The World Cities Culture Report* (Mayor of London, 2014).

Above, we have explored the cultural industries on a national level via GDP and employment. When one looks at the revenues of the urban cultural sector, depicted in Table 9.7, it appears that the developed urban economies perform better than the emerging urban economies. An exception hereto is Beijing, of which the revenues were higher than those of the cultural prominent city of Berlin. The shares of GDP generated by the urban cultural sectors of the three Chinese cities, shown in Table 9.7, are significantly higher than those of the developed urban economies, shown in the table.

If one looks at the share of cultural industries in total urban employment, there is one city located in an emerging economy which stands out: Mumbai, home of the Bollywood films. In 2005, no less than 16 per cent of the total labour force in Mumbai was employed in the cultural industries thereby surpassing the shares of established global cultural capitals like Los Angeles (12.5 per cent), London (11.3

Table 9.7 The total revenue of the urban cultural sector and the share of urban GDP generated by the urban cultural sector of the selected cities (Amsterdam: annually; Berlin: 2012; Sydney: 2012; Toronto: annually; Montreal: 2012; Los Angeles: 2014; Beijing: 2010; Hong Kong: 2013; Singapore: 2000; Taipei: 2011)

City	*Revenues of cultural sector in US dollars (billion)*	*Percentage of city's GDP generated*
Berlin	18.2	10
Sydney	N/A	9
Montreal	6.6	N/A
Los Angeles	82.6	N/A
Shanghai	N/A	11.3
Beijing	27.3	12.2
Hong Kong	17.1	5.1
Singapore	2.2	1.9
Taipei	12.2	N/A

Sources: Yingying *et al.*, 2013; Third Creative Industries Report, 2015; Census and Statistics Department, Hong Kong Special Administrative Region, 2015; Information Office of Shanghai Municipality, 2013; Mayor of London, 2014; Los Angeles County Economic Development Corporation, 2015; City of Sydney, 2015; Cho and Hoo 2003; Board of Trade of Metropolitan Montreal, 2013; Department of Economic Development, Taipei City Government Investment in Taipei Area, n.d.

per cent) and New York (8 per cent) (Mayor of London, 2014; Los Angeles County Economic Development Corporation, 2015; Bakhsi *et al.*, 2015 and Forman, 2015) and that of upcoming cultural capitals Seoul (9.4 per cent) and Shanghai (7.4 per cent). The shares of the cultural industries in total employment in the other cities located in emerging economies for which we have data are still way below these figures, i.e. São Paulo: 2.3 per cent and Rio de Janeiro: 2.2 per cent (Sistema Firjan, 2012).

The cultural industries within cities

Film

Los Angeles with Hollywood and Mumbai's Bollywood are renowned urban centres in the global film industry. When further exploring the film industry within the emerging economies, it appeared that those national film industries tend to concentrate in the large urban centres. Small and medium-sized production, distribution and financing enterprises, as well as large film conglomerates are to be found in the large urban centres like Seoul, Beijing, Moscow, St. Petersburg, and São Paulo (China Hollywood Society, n.d;. Korean Film Biz Zone, n.d.; Russianfilm, n.d.; https://en.wikipedia.org/wiki/Category:Film_production_companies_of_Brazil, n.d.)

Music

The contemporary global music industry is dominated by three large companies, Sony BMG Music, Warner Music Group and Universal Music Group, which together comprised over 70 per cent of the global music market in 2014 (Copyright Law, Treaties and Advice, 2007; Forden, 2012; Music and Copyright, 2015). All three enterprises have their headquarters located in New York, while Sony BMG, a merger between the American Sony company and the German firm Bertelsmann, has a second headquarter located in Berlin. The three large companies also have offices in the urban centres in emerging economies (e.g. Hong Kong, Johannesburg, Moscow, Mumbai, Rio de Janeiro, Seoul, Singapore, Taipei, Beijing, Shanghai and São Paulo) (Universal Music Publishing Group, n.d.; Sonyatv, n.d; BMG, n.d.; Warner/Chappell Music, n.d) which was also noted by Scott (2000: 208) who observed a scattering of dependent branch plants and affiliates over 'an enormous geographic area'.

Architecture

In a similar manner as uncovering the urban music industry, it is explored in which urban centres the ten largest architecture firms house their international offices. All the firms originating from developed economies, being the majority in the top-ten, have their headquarters located in cities within these developed economies like London, New York and Los Angeles. The three companies originating from either China or Singapore had most of their international offices located within Asia. The Chinese P&T Architects, even had no offices located in the United States or Europe. Furthermore, none of the ten firms had any offices on the African continent, and seven out of ten had no offices in South America (Aecom, n.d.; Nikken Sekkei, n.d.; Gensler, n.d.; Perkins & Will, n.d.; Aedas, n.d.; HOK, n.d.; Samoo Architects & Engineers, n.d.; IBI Group, n.d.; Fosters & Partners, n.d. and DP Architects, n.d.).

The trend of architecture being on the rise in Asia (Ren, 2011), is confirmed by our data on the location of architectural design. Furthermore, it corroborates the general observation that both Africa and South America lag behind in the shift to services-dominated economy by showing that they are also absent when it comes to housing globally renowned architecture firms.

Fashion

Paris, London, New York, and Milan, all located in developed economies (Jansson and Power, 2010: 891) are, without any doubt, major, if not the fashion capitals of the world. In these cities, the grand designer houses are located and very well-publicised biannual fashion weeks showcase both ready-to-wear fashion and haute couture. These renowned promotional events showcasing new fashions and new styles do not just rely on the designers located in these cities, but also on a dedicated physical infrastructure providing exhibition areas with spectacular show spaces to house the catwalks and a social infrastructure of ateliers able to actually turn the designs into clothes, make-up artists, sound and light engineers and other supporting staff (Jansson and Power, 2010). In addition, many promotional fashion events are managed by prominent agencies such as IMG Fashion and the Fédération Française de la Couture du Prêt-à-Porter des Couturiers and Créateurs de Mode which have the specific knowledge to organise them (IMG, n.d.; Mode a Paris, n.d.).

What we do see, however, is an increasing number of cities in emerging and BRICS economies holding fashion weeks of their own. Some of them are even organised by IMG Fashion (e.g. fashion weeks in Moscow and Mumbai, IMG, n.d.). These fashion weeks are typically sponsored by multinationals such as Mercedes Benz (Beijing, Moscow) or the Sunglasshut (Johannesburg).

Export of cultural industries

Thus far, the production of several cultural industries has been discussed on a national and urban scale. However, the cultural products and services produced are not solely consumed on the domestic market, but are increasingly exported to other countries. Table 9.8 shows the total value of the exported cultural goods and services from the emerging economies.

From Table 9.8 it can be concluded that China stands head and shoulders above the rest in terms of cultural export revenues. China's total 103 billion US dollars of cultural exports consisted mainly of the exports of creative goods such as

Table 9.8 The approximate cultural exports of the emerging economies in US dollars, when possible, divided in services and products

Country	*Cultural export (year) in US dollars (billion)*	*Exported cultural services in US dollars (billion)*	*Exported cultural products in US dollars (billion)*
China	103 (2010)	3	100
India	18 (2010)	4	14
Russia	8 (2010)	6	2
Brazil	7 (2010)	6	1
Singapore	2.7 (2000)	N/A	N/A
South Africa	0.35 (2010)	0	0.35

Source: CISAC, 2014.

garment design, arts crafts and new media (100 billion dollars). However, 70 per cent of these goods were produced by foreign enterprises, which produced solely for export. India and South Africa exported more cultural goods than services as well, whereas Brazil and Russia were the only BRICS countries exporting more cultural services – such as architecture design, engineering and advertising – than cultural goods (CISAC, 2014).

The cultural exports of the BRICS have grown significantly in recent years, but combined their cultural exports were still half of the cultural exports of the United States in 2010 (CISAC, 2014). However, the cultural exports of all developing countries, as defined by UNCTAD (2013) are half of the global cultural export in 2011 (UNCTAD, 2013).

Conclusion

Cultural industries are one of the mainstays of advanced economies and more in particular of advanced urban economies (Scott, 2012). These advanced urban economies are to an ever lesser extent the prerogative of Western countries. We now find cutting-edge economic activities – high-tech, producer services, research and development, and cultural industries in a whole array of cities across the world. The global division of labour is changing as new locations for advanced economic activities emerge. In this chapter, we have focused on one particular set of activities, namely the cultural industries. We have looked at the East Asian Tiger economies and the BRICS, countries which have already moved to (at least) an intermediate stage of economic development and are, hence, obvious candidates for becoming significant global cultural players. The available data are still rather incomplete and inconsistent, making a full-scale comparison as yet impossible. We can, however, present a preliminary mapping which allows a few tentative conclusions.

First, our data show that the United States and the United Kingdom are still the frontrunners when measured in absolute revenues. Secondly, when one would use the (relative) size of cultural industries as a yardstick for the stage of economic development, then Brazil and Taiwan are clearly catching up. The relative contribution to GDP, as our data shows, is in the same league as that of the United States. A third finding is that if one zooms in on specific sectors and cities, the picture becomes more nuanced. Film from India, music from Brazil and South Korea, and architectural design from China are examples of upcoming cultural industries in emerging economies not just catering to domestic but also foreign markets. Fashion design, however, is still dominated by the established global brands from Western countries. A fourth finding concerns the role of cities. Data on individual cities in emerging economies are even harder to come by, but cities like Mumbai, with no less than 16 per cent of total formal employment in cultural industries, Beijing, Singapore, Hong Kong, Seoul, São Paulo and Rio Janeiro are becoming hot spots of cultural industries.

These findings are a very much in line with the UNCTAD reports (2008, 2010, 2013) and Allen Scott's expectation (2000: 209): 'Provided that the right mix of entrepreneurial know-how, creative energy, and public policy can be brought to bear on the relevant developmental issues, there is little reason why these cities cannot parlay their existing and latent cultural-products sectors into major global industries.'

The view that cultural industries may be one of the drivers of the next stage of economic development is also spreading among policy makers in emerging economies. Already in the 1990s, policy makers in South Korea embarked on a set of conscious policies to build up export-oriented cultural industries. These policies were clearly based on the successful template used earlier to foster manufacturing (e.g. Samsung, Hyundai and Daewoo) by stimulating export (Shim, 2006). In a sense, then, Psy's and more in general K-Pop's success could be seen as the next stage of South Korea's economic development. The government of Singapore (Kong, 2012) and even the Communist Party of China (Kong *et al.*, 2015) have now officially identified cultural industries as policy targets.

These policies typically start with import substitution, often to counterbalance the American cultural dominance (Osterhammel and Peterson, 2003). The domestic market is, however, just one phase and the next step is to export films, music, and other cultural products and services. Usually export is directed towards cultural and/or linguistic proximate regional markets like India has done with films in East Asia, South Korea with music and soaps in Asia, and Brazil also with music and soaps in Latin America. However, the ultimate step towards global markets might be more difficult given the difference between cultural realms and the strength of Western media companies and systems houses more in general controlling access to global networks and markets. In addition, it is questionable whether these countries do possess an open, tolerant cosmopolitan cultural climate that is essential for cultural innovations. The fact that Psy's single 'Right Now' was banned from under-19 audiences by South Korea's Ministry of Gender Equality and Family for its 'obscene' lyrics sheds a doubtful light on the tolerance regarding creativity (https://en.wikipedia.org/wiki/Psy).

Note

1 According to some statistics, the Nigerian film industry comes after India in terms of films produced, the Nigerian films, however, are often semi-professionally produced (direct-to-video), and not released in theatres, (UIS, 2013).

References

Aecom (n.d.). *Offices & Contacts.* Locator of International Offices [Online] Available at: www.aecom.com/Offices+and+Contacts/Offices [accessed 23 July 2015].

Aedas (n.d.). *Where we are/Contact us.* Locator of International Offices [Online] Available at: www.aedas.com/en/contact [accessed 23 July 2015].

Appadurai, A. (1990). Disjuncture and difference in the global cultural economy. *Theory, Culture And Society*, 7(2), pp. 295–310.

Bakhsi, H., Davies, J., Freeman, A. and Higgs, P. (2015). *The Geography of the UK's Creative and High-Tech Economies.* London: Nesta.

Bautès, N. and E. Valette (2004). Miniature painting, cultural economy and territorial dynamics in Rajasthan, India. In: D. Power and A.J. Scott (eds), *Cultural Industries and the Production of Culture.* London: Routledge, pp. 207–23.

BMG (n.d.). *International Presence.* Locator of International Offices [Online] Available at: www.bmg.com/category/about-us/international-presence/ [accessed 3 August 2015].

Brandellero, A. and Pfeffer, K. (2011). Multiple and Shifting Geographies of World Music Production, *Area,* 43(4), pp. 495–505.

Board of Trade of Metropolitan Montreal (2013). The Creative Industries: Catalysts of Wealth and Influence for Metropolitan Montreal, Arts Business, November 2013.

Boxofficeindia (2015). *Bajrangi Bhaijaan Closing In On 500 Crore Worldwide.* Online article on the gross worldwide business of Indian films [Online] Available at: www.boxofficeindia.com/Details/art_detail/bajrangibhaijaanclosinginon500crore-worldwide#.Vkc6CfkvfWK [accessed 15 August 2015].

Boxofficemojo (n.d.). *All time box office.* Listing of Worldwide Grosses [Online] Available at: www.boxofficemojo.com/alltime/world/ [accessed 15 August 2015].

By Koket (2013). *Top 50 Best Fashion Designers.* Love Happens Blog; By Koket. [Online] Available at: www.bykoket.com/blog/top-50-fashion-designers/ [accessed 17 August 2015].

Caldwell, C. (2012). Gangnam stylishly debunks US myth, *Financial Times*, December 28th, p. 7.

Census and Statistics Department, Hong Kong Special Administrative Region (2015). *The Cultural and Creative Industries in Hong Kong.* Hong Kong Monthly Digest of Statistics June 2015. Hong Kong: Census and Statistics Department.

CEOworld Magazine (2011). *List of World's Top 20 Fashion Designers – Best Fashion Designers in the World.* CEOworld Magazine. [Online] Available at: http://ceoworld.biz/ceo/2011/11/29/list-of-worlds-top-20-fashion-designers-best-fashion-designers-in-the-world [accessed 17 August 2015].

Chang, H.J. (2002). *Kicking Away the Ladder: Development strategy in historical perspective.* London: Anthem Press.

China Hollywood Society (n.d.). *Chinese Film Studios* [Online] Available at: www.chinahollywood.org/chinese-film-studios [accessed 24 July 2015].

Cho, A. and Hoo, T. (2003). *Economic Contributions of Singapore's Creative Industries.* Economic Survey of Singapore First Quarter 2003.

CISAC (2014). *The Creative Industries and the BRICS; A review of the state of the creative economy in Brazil, Russia, India, China and South Africa*. Confédération Internationale des Sociétés d'Auteurs et Compositeurs.

City of Sydney (2015). *The creative Economy*. [Online] Available at: www.cityofsydney.nsw.gov.au/business/business-support/business-in-your-sector/key-industries/creative-and-digital [accessed 29 July 2015].

Copyright Law, Treaties and Advice (2007). *The Big Four Record Companies*. [Online] Available at: www.copynot.org/Pages/The%20big%20four%20Record%20 Companies.html [accessed 31 July 2015].

Cultural and Creative Industry Promotion Team (2004). *The Development of the Cultural and Creative Industries in Taiwan and Its Significance for SMEs*. Taipei: Ministry of Economic Affairs.

Currid, E. (2007). *The Warhol Economy: How Fashion, Art and Music Drive New York City*. Princeton: Princeton University Press.

Currid, E., and Williams, S. (2010). Two cities, five industries: similarities and differences within and between cultural industries in New York and Los Angeles, *Journal of Planning Education and Research,* 29(3), pp. 322–335.

DCMS (2015). *Creative Industries Economic Estimates January 2015; Statistical Release*. London: Department for Culture Media & Sport.

Department of Economic Development, Taipei City Government Investment in Taipei Area (n.d.). *Creative Economy: Taipei Has Its Thinking Cap On*. [Online] Available at: www.doed.gov.taipei/ct.asp?xItem=79408820&ctNode=73050&mp=10500W [accessed 7 August 2015].

Dixon, C. (2009). East Asian Miracle. In: R. Kitchen and N. Thrift, eds., *International Encyclopedia of Human Geography, 1st Edition*. Amsterdam: Elsevier, pp. 273–79.

DP Architects (n.d.). *Contact*. Locator of International Offices [Online] Available at: www.dpa.com.sg/contact/#china [accessed 23 July 2015].

Economist, The (2012). "Hot spots: Benchmarking global city competitiveness", Economist Intelligence Unit, www.citigroup.com/citi/citiforcities/pdfs/hotspots.pdf [accessed on 19 June 2015].

Economist, The (2014). "What China wants", 23 August, pp. 37–42

Erkan (2013). *10 Most Famous Fashion Designers of All Time*. The Most 10. [Online] Available at: www.themost10.com/famous-fashion-designers-all-time/ [accessed 17 August 2015].

Ford, E. (n.d.). *Five famous architects and their most iconic buildings*. Architecture & Design. [Online] Available at: www.architectureanddesign.com.au/features/features-articles/five-famous-architects-and-their-most-iconic-build [accessed 19 August 2015].

Forden, S. (2012). *Universal Music Wins Approval to Buy EMI Recorded Music*. Online article: Bloomberg [Online] Available at: www.bloomberg.com/news/articles/2012-09-21/ftc-closes-review-of-universal-music-groups-emi-purchase [accessed 31 July 2015].

Forman, A. (2015). *Creative New York*. New York: Centre for Urban Future.

Fosters & Partners (n.d.). *Contact Us/Offices*. Locator of International Offices [Online] Available at: www.fosterandpartners.com/contact/offices [accessed on 23 July 2015].

Frieden, J. (2006). *Global Capitalism, Its Fall and Rise in the Twentieth Century*. New York/London: W.W. Norton and Company.

Garrahan, M. (2015). *Will Streaming Save the Music Industry?* Financial Times. [Online] Available at: www.ft.com/intl/cms/s/2/005166f6-ac6e-11e4-9d32-00144feab7de.html# axzz3jjAscj8X [accessed 19 August 2015].

Gensler (n.d.). *Offices*. Locator of International Offices [Online] Available at: www.gensler.com/offices [accessed 23 July 2015].

Gibson, R. (2012). *Iconic Legends: The 10 Greatest Modern Architects of Our Time*. Freshome; Designs & Architecture Magazine [Online] Available at: http://freshome.com/2012/08/17/iconic-legends-the-10-greatest-modern-architects-of-our-time [accessed 19 August].

Hall, P. (1998). *Cities in Civilisation: Culture, Innovation, and Urban Order*. London: Weidenfeld & Nicolson.

Henderson, K. (2012). *25 Architects You Should Know*. Complex [Online] Available at: http://uk.complex.com/style/2012/05/25-architects-you-should-know/ [accessed 19 August 2015].

HOK (n.d.). *Locations*. Locator of International Offices [Online]. Available at: www.hok.com/about/locations [accessed on 23 July 2015].

Hoyler, M. and Watson, A. (2012). Global media cities in transnational media networks. *Tijdschrift voor economische en sociale geografie*, 104(1), pp. 90–108.

Hunter Tilney, L. (2015). *How to Succeed in the Music Industry? Financial Times* [Online]. Available at: www.ft.com/intl/cms/s/0/a050da64-0471-11e5-95ad-00144feabdc0.html [accessed 19 August 2015].

IBI Group (n.d.). *Contact Us*. Locator of International Offices [Online] Available at: www.ibigroup.com/contact [accessed 23 July 2015].

IMG (n.d.). *IMG-Fashion*. [Online] Available at: http://img.com/services/event-management/fashion.aspx [accessed 8 August 2015].

Information Office of Shanghai Municipality (2013). *Development of Shanghai's cultural and creative industries*. Press Conference on 27 June 2013, Shanghai: Shanghai Municipal Government Information Office. [Online] Available at: http://en.shio.gov.cn/presscon/2013/07/12/1152451.html [accessed 20 July 2015].

International Business Times (2011). *World's Top 10 Most Popular Fashion Designers*. International Business Times. [Online] Available at: www.ibtimes.com/worlds-top-10-most-popular-fashion-designers-552189 [accessed 17 August 2015].

Jansson, J. and Power, D. (2010). Fashioning a global city: global city brand channels in the fashion and design industries, *Regional Studies*, 44(7), pp. 889–904.

Keane, M. (2004). Brave new world: Understanding China's creative vision, *International Journal of Cultural Policy*, 10(3), pp. 265–79.

Keane, M., Ryan, M.D. and Cunningham, S. (2005). Worlds apart? Finance and investment in creative industries in the People's Republic of China and Latin America, *Telematics and Informatics*, 22(4), pp. 309–331.

Kloosterman, R.C. (2004). Recent employment trends in the cultural industries in Amsterdam, Rotterdam, The Hague and Utrecht; a first exploration, *Tijdschrift voor economische en sociale geografie,* 95(2), pp. 243–62.

Kloosterman, R.C. (2008). Walls and bridges: knowledge spillover between 'superdutch' architectural firms, *Journal of Economic Geography,* 8(4), pp. 545–63.

Kloosterman, R.C. (2010). Building a Career: Labour Practices and Cluster Reproduction in Dutch Architectural Design, *Regional Studies,* 44(7), pp. 859–71.

Kloosterman, R.C. (2015). Forces of agglomeration: Allen Scott's The Cultural Economy of Cities revisited, *Built Environment,* 41(3), pp. 379–89.

Kloosterman, R.C., Beerepoot, N. and Lambregts, B. (2015). Service-sector driven economic development from a historical perspective. In: Lambregts, B., Beerepoot, N., and Kloosterman, R.C. (eds), *The Local Impact of Globalization in South and Southeast Asia: Offshore Business Processes in Services Industries*. Oxford/New York:

Routledge, pp. 17–28.
Kong, L. (2012). Ambitions of a global city: arts, culture and creative economy in 'Post-Crisis' Singapore, *International Journal of Cultural Policy*, 1(3), pp. 279–94.
Kong, L., Chia-ho, C. and Tsu-Lung, C. (2015). *Arts, Culture and the Making of Global Cities: Creating New Urban Landscapes in Asia*. Cheltenham: Edward Elgar Publishing.
Korean Film Biz Zone (n.d.). *Companies Directory.* [Online] Available at: www.koreanfilm.or.kr/jsp/films/index/companyList.jsp?pageIndex=1 [accessed 24 July 2015].
Krätke, S. (2003). Global media cities in a world-wide urban network. *European Planning Studies,* 11(6), pp. 605–28.
Kwon, S. and Kim, J. (2013). 'The cultural industry policies of the Korean government and the Korean Wave', *International Journal of Cultural Policy*, 20(4), pp. 422–39
Los Angeles County Economic Development Corporation (2015). *2014 Otis Report on the Creative Economy of the Los Angeles Region*. Los Angeles: Los Angeles County Economic Development Corporation.
Mayor of London (2014). *World Cities Culture Report*. London: Greater London Authority.
McNeill, D. (2009). *The Global Architect; Firms, Fame and Urban Form*. NewYork/London: Routledge.
Mode a Paris (n.d.). *Fédération Française de la Couture du Prêt-à-Porter des Couturiers and Créateurs de Mode* [Online] Available at: www.modeaparis.com/en [accessed 27 July 2015].
Music and Copyright (2015). *Recorded music market share gains for WMG in 2014, Sony/ATV is the publishing leader*. Blogpost on Music & Copyright's Blog. [Online] Available at: https://musicandcopyright.wordpress.com/tag/market-share [accessed 31 July 2015].
National Endowment for the Arts (2013). *U.S. Bureau of Economic Analysis and National Endowment for the Arts Release Preliminary Report on Impact of Arts and Culture on U.S. Economy*. [Online] Available at: http://arts.gov/news/2013/us-bureau-economic-analysis-and-national-endowment-arts-release-preliminary-report-impact [accessed 3 August 2015].
Nikken Sekkei (n.d.). *Contacts*. Locator of International Offices [Online] Available at: www.nikken.co.jp/en/contacts/index.html [accessed 23 July 2015].
Nye J.S. (1990). Soft Power. *Foreign Policy,* 80, pp. 153–71.
O'Connor, J. (2004). Cities and culture and "transitional economies": developing cultural industries in St. Petersburg. In: Power, D. and Scott, A.J. (eds), *Cultural Industries and the Production of Culture*. London: Routledge, pp. 37–53.
O'Neill, J. (2011). *The Growth Map: Economic opportunity in the BRICS and beyond.* London: Penguin UK.
Osterhammel, J. and Peterson, N.P. (2003). *Globalization; A Short History*. Princeton/Oxford: Princeton University Press.
Perkins & Will (n.d.). *Locations*. Locator of International Offices [Online]. Available at: http://perkinswill.com/ [accessed 23 July 2015].
Power, D. and Hallencreutz, D. (2002). Profiting from Creativity? The Music Industry in Stockholm, Sweden and Kingston, Jamaica, *Environment and Planning A*, 34(10), pp. 1833–54.
Pratt, A. C. (1997). 'The cultural industries production system: a case study of employment change in Britain, 1984–91', *Environment and Planning A*, 29(11), pp. 1953–1974.
Pratt, A. C. (2012). The cultural economy and the global city. In: Derudder, B., Hoyler, M.,

Taylor, P.J. and Witlox, F. (eds), *International Handbook of Globalization and World Cities*. Cheltenham: Edward Elgar, pp. 265–74.
Pratt, A. C. (2013). The cultural and creative industries: Organisational and spatial challenges to their governance, *DIE ERDE*, 143(4), pp. 317–34.
Ren, X. (2011). *Building Globalization; Transnational Architecture Production in Urban China*. Chicago/London: University of Chicago Press.
Ren, X. and Sun, M. (2012). Artistic urbanization: creative industries and creative control in Beijing, *International Journal of Urban and Regional Research*, 36(3), pp. 504–21.
Russianfilm (n.d.). *Film Making in Russia*. [Online] Available at: http://russianfilm.biz/companies/ [accessed 24 July 2015].
Samoo Architects & Engineers (n.d.). *Contact*. Locator of International Offices [Online] Available at: www.samoo.com [accessed on 23 July 2015].
Sandbrook, D. (2015). *The Great British Dream Factory: The Strange History of Our National Imagination*. London: Allen Lane.
Sassen, S. (1991/2001). *The Global City: New York, London, Tokyo*. Princeton: Princeton University Press.
Scott, A. J. (2000). *The Cultural Economy of Cities: Essays on the Geography of Image-Producing Industries*. London: Sage.
Scott, A. J. (2004). Cultural-products industries and urban economic development: Prospects for growth and market contestation in global context, *Urban Affairs Review*, 39(4), pp. 461–90
Scott, A.J. (2008). *Social Economy of the Metropolis: Cognitive-Cultural Capitalism and the Global Resurgence of Cities*. Oxford: Oxford University Press.
Scott, A.J. (2012). *A World in Emergence, Cities and Regions in the 21st Century*. Cheltenham: Edward Elgar.
Shim, D. (2006). Hybridity and the rise of Korean popular culture in Asia. *Media, Culture & Society*, 28(1), pp. 25–44.
Sistema Firjan (2012). *Indústria Criativa; Mapeamento da Indústria Criativa no Brasil*. Sistema Firjan; Informa, Forma, Transforma.
Sonyatv (n.d.). *Languages*. Locator of International Offices [Online] Available at: www.sonyatv.com/languages.php [accessed 31 July 2015].
Statista (2015). *Brand value of the leading 10 apparel brands worldwide in 2015 (in million U.S. dollars)*. [Online] Available at: www.statista.com/statistics/267931/brand-value-of-the-leading-10-apparel-brands-worldwide [accessed 7 August 2015].
Statisticbrain (2015). *Fashion Industry Statistics*. [Online] Available at: www.statisticbrain.com/fashion-industry-statistics [accessed 7 August 2015].
Storper, M. (1997). *The Regional World: Territorial Development in a Global Economy*. New York: Guilford Press.
Straubhaar, J. (2010). Chindia in the context of emerging cultural and media powers. *Global Media and Communication*, 6(3), p. 253.
The Recording Industry Association of Japan (2015). *Statistics Trends; The Recording Industry in Japan 2015*. RIAJ Yearbook 2015. Tokyo: Recording Industry Association of Japan.
Third Creative Industries Report (2015). *Third Creative Industries Report; Development and Potential*. Berlin: Senatsverwaltung für Wirtschaft, Technologie und Forschung.
Thussu, D. K. (2013). India in the international media sphere, *Media, Culture & Society*, 35(1), pp. 156–62.
UNCTAD (2008). *Creative Economy Report 2008*, United Nations Conference on Trade and Development, 2008, Retrieved from http://unctad.org/en/Docs/

ditc20082cer_en.pdf [accessed on 28 August 2015].
UNCTAD (2010). *Creative Economy Report 2010*, United Nations Conference on Trade and Development, Retrieved from http://unctad.org/en/docs/ditctab20103_en.pdf [accessed on 28 August 2015].
UNCTAD (2013). *Creative Economy Report 2013. Special Edition: Widening Local Development Pathways,* United Nations Conference on Trade and Development. Retrieved from www.unesco.org/culture/pdf/creative-economy-report-2013.pdf [accessed on 4 July 2015].
UNESCO Institute for Statistics (2013). *Emerging Markets and the Digitalization of the Film Industry: An Analysis of the 2012 UIS International Survey of Feature Film Statistics.* UIS Information Paper No. 14. Montreal: UNESCO Institute for Statistics.
Universal Music Publishing Group (n.d.). *Global Contacts*. Locator of International Offices [Online] Available at: www.umusicpub.com/#contentRequest=globalcontacts&contentLocation=&contentOptions= [accessed 31 July 2015].
WA100 (2015). The big list Features Building Design. 2015. *WA100 2015: The big list Features Building Design*. [Online] Available at: www.bdonline.co.uk/wa100-2015-the-big-list/5072847.article [accessed 5 August 2015].
Wallis Simons, J. (2014). *World Architecture Festival 2014: Vietnamese firm wins best building award.* CNN. [Online] Available at: http://edition.cnn.com/2014/10/03/travel/world-architecture-festival-2014 [accessed 19 August 2015].
Warner/Chappell Music (n.d.). *Global Offices*. Locator of International Offices [Online] Available at: www.warnerchappell.com/about [accessed 3 August 2015].
Watson, A. (2012). The world according to iTunes: mapping urban networks of music production. *Global Networks*, 12(4), 446–66.
Wikipedia (n.d.). *Film Production Companies of Brazil*. [Online] Available at: https://en.wikipedia.org/wiki/Category:Film_production_companies_of_Brazil [accessed 11 August 2015].
Wikipedia (n.d.). *Gagnam Style*. [Online] Available at: https://en.wikipedia.org/wiki/Gangnam_Style [accessed 15 November 2015].
Wikipedia (n.d.). *Psy.* [Online] Available at: https://en.wikipedia.org/wiki/Psy [accessed on 15 November 2015].
Yingying, X., Duo, J., McArthur, I. and Miller, B. (2012). Shaping cultural and creative space: Beijing as a case study. In: Cleland, K., Fisher, L. and Harley, R. (eds), *Proceedings of the 19th International Symposium of Electronic Art*. Sydney, pp. 1–4.
Yusuf, S. and Nabeshima, K. (2005). Creative industries in East Asia, *Cities,* 22(2), pp. 109–122.

Part III

Service-led modernisation in the Global South

10 Changing geographies of service delivery in South Africa

Towards regional value chains?

Charlotte Keijser

Introduction

Global value chains (GVCs) are a defining feature of the current wave of globalisation and a reflection of the increasing interconnectedness of economies (OECD, 2015). Many firms, predominantly from more advanced economies, seek efficiency gains and cost savings by delegating their non-core activities to external firms in the country (that is outsourcing) or to firms abroad, mostly in developing countries (offshore outsourcing or offshoring) (Gereffi and Fernandez-Stark, 2010). This has given rise to value chains of services, typically referring to business processes or functions with usage of information technology (IT) in the delivery process, which is also known as the IT-enabled services (ITES) industry. These value chains of IT-enabled services include a broad spectrum of services, ranging from customer services and data processing to higher value-added work, such as software development, risk management, and research and design (Dossani and Kenney, 2007; Palugod and Palugod, 2011).

The ITES industry has opened up opportunities for developing countries to participate in GVCs of services without having to develop a full range of functional capabilities. Participation in GVCs of services is argued to be vital for their economic development in terms of diversification and sophistication of exports, access to markets, and considered an opportunity to add value to their local industries (Gereffi *et al.*, 2010; OECD, 2015). Understanding how developing countries can successfully integrate into GVCs of services has thus become an important subject for development practitioners.

South Africa is an interesting case, as the country only in recent years made name for itself as a provider of IT-enabled services like customer relations, data management and legal services. Furthermore, the case of South Africa shows that it does not solely engage in GVCs through acting as a delivery platform for services from overseas, but also as an end market for IT-enabled services, and as a gateway to the rest of Africa. This has attracted foreign service providers to South Africa seeking ITES contracts in the domestic and sub-Saharan Africa (SSA) market, giving rise to local value chains (LVCs) and regional value chains (RVCs) of services. The resulting heterogeneity of value chains is partly driven by firms' strategies to create a global delivery model, which refers to having a global network of service delivery centres (Gereffi *et al.*, 2010; Sako, 2009). In addition, this study reveals that some South African owned service providers struggle to compete on the global market and have instead turned to the domestic and regional ITES market. This is also evident in African agriculture where policy measures have already been directed at developing and promoting RVCs through harmonising national policies, laws and regulation, and standards across regional economic communities in SSA (Ramdoo, 2014). The key argument is that RVCs have the potential of expanding market opportunities, especially for African countries with small populations and fragmented national markets. National and regional markets in developing countries offer an alternative to the global market that is characterized by strong competition, high-quality standards and that favours large-scale producers (ERA, 2009).

The South African ITES industry fits well into recent academic debates on the need to look at the potential of RVCs of services, defined as chains of activities where the final product (here a service) is provided from a country within the region (here the SSA market). This subject is especially important, as entering GVCs of services is vital for economic development, yet often challenging for firms in developing countries. Scholars argue that RVCs offer them more opportunities to participate, gain experience and build capabilities needed to compete globally, and thus serve as a stepping stone into GVCs (see Bamber *et al.*, 2013).

While RVCs have captured increasing interest by researchers and policymakers, empirical evidence mainly remains limited to manufacturing sectors. Drawing on primary and secondary data collected in South Africa this chapter analyses heterogeneous value chains of IT-enabled services. It particularly sheds light on what RVCs of services look like, in terms of how services are organized in the regional market through different sourcing modes. In doing so, this chapter will provide a conceptual background (section 2), methodology (section 3), followed by an overview of South Africa's ITES industry (section 4), and examples of RVCs leading to an extension of modes of outsourcing/offshoring as currently reflected in the literature (section 5 and 6).

Theoretical overview

Outsourcing and offshoring

The globalisation of production and trade are important features of the contemporary economy (Gereffi *et al.*, 2001). World markets are rapidly integrating while at the same time production is becoming more fragmented through lower transportation costs and technological advances. The term outsourcing is used to refer to a firm's decision to not undertake all of its business functions internally, but to outsource to an external firm. Offshoring involves a transfer of business activities that had been previously performed in the home country to another country; independent of whether simultaneous outsourcing (transfer of ownership of the activity) is occurring. In doing so, business functions can be performed offshore within the same firm (captive or subsidiary offshoring) or by an independent service provider (offshore outsourcing) (see Table 10.1) (Massini and Miozzo, 2012). The contracting of services to a firm in a nearby country is referred to as nearshoring.

According to Massini and Miozzo (2012), the current wave of outsourcing and offshoring concerns the relocation of service activities from large global operators and less internationalised companies from developed to less developed countries (see also Gereffi, 2005; Gereffi *et al.*, 2010; Oshri *et al.*, 2015). Firms set up captive units offshore in lower cost countries to provide services for the parent country (offshoring) or outsource it to an external service provider offshore (offshore outsourcing). As such, a range of activities that were once considered strictly the domain of the industrialised world are now also performed in developing countries (Gereffi *et al.*, 2011). Furthermore, it has opened up opportunities for developing countries to engage in international trade of services without having to develop the full range of capabilities, but only to specialise in specific stages of the chain. According to Sturgeon and Memedovic (2010) through participating in GVCs of services, developing countries can leap-frog development and make non-linear catch up possible.

Global and regional value chains

Global value chain scholars have devoted several studies to the act of outsourcing and offshoring, particularly how developing countries can insert themselves successfully into value chains of services (see Bamber *et al.*, 2013; Gereffi, 2005; Gereffi *et al.*, 2010). Although the value chain from production to consumption of services is commonly short, the GVC approach has proven useful to understand shifting patterns of global production, capabilities key for competitiveness, and the role played by different actors in developed and less developed countries (Gereffi *et al.*, 2010).

Table 10.1 Sourcing modes of services by client firms

		Location of client	
		Home country (Domestic market)	*Overseas (Global market)*
Type of provider	External (independent service provider)	*Domestic outsourcing* Outsourcing of services to an independent provider in the firm's home country	*Offshore outsourcing* Outsourcing of services to an independent provider abroad
	Internal (captive unit)	*Domestic captive* Service delivery is performed at home by the firm itself	*Offshore captive* Firm sets up affiliate abroad to provide in-house services

Source: Adapted from Massini and Miozzo, 2012.

For the purpose of this study, the terminology of Bamber *et al.*, (2013) is used to refer to RVCs as chains of activities at the regional market level and GVCs encompassing chains at the global market level, that is, across two or more continents. RVCs differ from GVCs in that the end product (here a service) of a RVC is provided by a country within the region, often within trading blocs; rather than across continents. However, questions remain concerning the definition of these so-called RVCs and how they differ from GVCs, for instance in terms of key drivers and (lead) actors. Furthermore, limited empirical evidence is available on what these regional chains look like in the context of services.

The case of sub-Saharan Africa

For many sub-Saharan African countries competing in the global market is a struggle, and their participation in GVCs is hampered by structural factors (OECD, 2015). It is argued that firms in these countries could gain experience and build the capabilities needed to compete globally by first reaping the benefits of intra-regional trade. The African regional market would function as a playing field where firms could 'learn by doing' and develop the capabilities and gain the experience needed to compete in global markets (OECD, 2015). Kaplinsky and Farooki (2010) argue that understanding consumer preferences in less advanced markets may be easier for firms from these markets or region, due to cultural, geographic and regulatory proximity. In addition, they claim that these markets often have less demanding standards compared to markets in advanced countries with respect to quality and product sophistication and often place more emphasis on price competitiveness.

In the context of outsourcing and offshoring of services, empirical literature on the opportunities of the regional market remains scant, especially with regard to studies on SSA. Only in recent years studies have been conducted on sub-Saharan African countries as new players in the IT-enabled services landscape, particularly on hub destinations Ghana, Kenya and South Africa (see Adelakun and Iyamu, 2013; Beerepoot and Keijser, 2015; Dihel, 2011; Mann and Graham, 2016; Wausi *et al.*, 2013). There is much common ground in these studies, as they all predominantly use a GVC perspective for analysing the 'offshore readiness' of these countries with regard to obtaining services contracts overseas, leaving little focus on the opportunities of the local or regional market.

Methodology

The aim of this chapter is to provide an overview of South Africa's ITES industry, explore different value chains of IT-enabled services, and to understand what sourcing modes give rise to RVCs of services. In doing so, this study draws upon academic papers, company data, industry reports, and other secondary sources. In addition, qualitative field research for this study was carried out in Cape Town from January until March 2015. Semi-structured interviews were conducted with CEOs and managerial staff of eight service providers based in Cape Town, of which the majority has multiple offices across South Africa. These companies were purposively selected to ensure that a comparison could be made between firms servicing the South African, regional and global market as well as a distinction between South African versus foreign owned companies. Short semi-structured interviews and panel discussions during the Business Processing enabling South Africa (BPeSA) Offshoring Summit (March 4–6, 2015), which brought together stakeholders, service operators and academia, led to information from six more firms.

On the institutional side, nine semi-structured interviews were conducted with representatives from government agencies, business associations and international organisations. These interviews aimed at understanding the ITES industry in terms of expectations, challenges and perceptions towards the opportunities on the local, regional and global market. The qualitative data gathered through semi-structured interviews and from policy reports and secondary data, were analysed using software programme Nvivo.

Analysis of South Africa's ITES industry

Industry overview

The beginning of South Africa's presence in the offshore market dates back to 1998 when Lufthansa was the first global company to choose Cape Town as a base for the delivery of its services. However, it was not until the year 2012 that South Africa experienced strong demand for ITES service delivery, especially in the offshore market reaching growth rates above 12 per cent per annum (Willcocks *et al.*, 2015). Despite recent growth, the ITES industry in South Africa is relatively small with around 200,000 jobs created, compared to approximately a million people working in the outsourcing/offshoring industry in the Philippines (BPAP, 2015) and approximately 3.5 million in India (NASSCOM, 2015). Domestic demand for captive and outsourcing operations accounts for the largest number of employees (see Table 10.2).[1] The domestic market is serviced both by local owned firms as well as foreign service providers mainly based in Johannesburg.

The offshore market is small, but has shown significant growth in recent years expanding from approximately 10,000 jobs in 2012 to over 25,000 jobs in 2015 (BPeSA, 2015). It concerns mainly firms from the United Kingdom that have offshored services to South Africa, due to historical links, family ties, and similarities in business environment, laws and regulations. The offshore market is serviced by captive units of parent firms or by independent (foreign) service providers (BPeSA, 2015).

International companies currently offshoring to South Africa include Amazon, American Airlines, British Gas, IBM, Lufthansa, Shell, Shop Direct and TalkTalk (BPeSA, 2015). The arrival of captive units of these international firms and expansion of existing operations of foreign service providers (e.g. Accenture, IBM and Infosys) have put the country on the map as player in the global ITES industry.

Table 10.2 Number of jobs created in the South African IT-enabled services industry by January 2015

Market	*Type of investment*		
	Captive	*Outsource*	*Total*
Domestic	64,056	124,344	188,400
Offshore	7,156	18,578	25,734
Total	71,212	142,922	214,134

Note: Domestic captives refer to 100 per cent subsidiaries of firms in South Africa, which provide services exclusively to the parent company. Offshore captives refer to 100 per cent subsidiaries of companies based outside South Africa, which provide services exclusively to the parent company outside South Africa.

Source: BPeSA, 2015.

Services are performed mainly in three cities, Johannesburg and the Gauteng region, Durban and the KwaZulu Natal (KZN) region, and Cape Town and the Western Cape region. These three main areas comprise what Willcocks *et al.*, (2015) refer to as the 'service triad'. Cape Town counts the largest number of customer related jobs coming from the offshore market, mainly due to the quality of its infrastructure and attractive expat lifestyle (Willcocks *et al.*, 2015). Service providers in Johannesburg and the wider Gauteng region predominantly focus on customer relations, IT and financial services in the domestic ITES market. This is because the region is home to most of the headquarters of multinational companies (MNCs) operating in Southern and Eastern Africa, and the centre of South Africa's main industries and financial centres. In the last decade firms in Durban and the wider KZN region have started to provide sales-related services to client firms in both the domestic and offshore market. The overall cost of doing business in KZN is lower than in the Gauteng and Western Cape region, and is generally home to providers opening up a second office in South Africa (Willcocks *et al.*, 2015).

Breakdown of ITES value chains

Service providers in South Africa deliver non-primary business process outsourcing (BPO) services as well as higher value-added services in the segments of Information Technology Outsourcing (ITO) and Knowledge Process Outsourcing (KPO) (see Table 10.3). The BPO segment includes low value-added services that do not require specialised skills and have minimum training requirements and qualifications compared to ITO and KPO activities (Gereffi *et al.*, 2010). The BPO segment is by far the largest segment, which consists for 70 per cent of front office services (for example customer relations such as call centre services), and the remaining of back-office services (like data processing). The ITO segment constitutes activities, such as coding, testing and maintenance of software and higher value-added services including software development and system integration. No accurate numbers are available, but it is argued that the ITO market is

Table 10.3 Different segments within value chains of IT-enabled services in South Africa

Information technology outsourcing (ITO)	*Knowledge process outsourcing (KPO)*	
IT consulting	Business consulting and research	Legal services/LPO
	Actuarial servicing/	Legal research
	Risk management	Compliance
	Accountancy	Contract management
Software		Mergers, acquisitions and due diligence review
Software development		Document and Contract review
System integration		Document drafting
Desktop management		
Infrastructure	*Business process outsourcing (BPO)*	
IT management	Front office/contact centres	Back office/data processing
IT infrastructure solutions	Marketing and Sales	HR/Payroll management
IT support	Credit collection	Transaction processing
	Customer services	Asset management
		Administration

Source: different secondary sources.

significantly smaller than the BPO market, due to fewer IT skills available (BPeSA, 2015). KPO entails high-end knowledge work, such as consulting practices and marketing research, and makes up the high-end of the value chain. Particularly, Legal Process Outsourcing (LPO) has been identified as a niche where South Africa can compete globally (Willcocks *et al.*, 2015). The LPO sector in South Africa is fairly new, but is fast growing and employs approximately 1,350 employees (Zille, 2015). South Africa has a large group of law graduates that have not made it to practice yet, while possessing the appropriate skills for LPO (for example paralegal skills). The LPO market in South Africa consists of standardised jobs such as document review and higher value-added services such as legal research.

Table 10.3 gives an overview of the different segments in the ITES sector, the financial industry is responsible for the highest demand for services, coming mainly from the domestic market, whereas the demand for services from offshore markets is mainly driven by the telecommunication industry (BPeSA, 2015). Capabilities are particularly built in both front and back-office services across several sub-verticals, such as insurance, banking, and asset management (Everest Group and Letsema Consulting, 2008).

Interviews reveal that upgrading along the value chain from BPO activities to higher value-added activities is considered crucial for the development of the industry, especially since its domestic BPO market is argued to be maturing. Furthermore, firms want more complex jobs for their employees to promote an attractive career path. Even though BPO remains the backbone of South Africa's ITES industry, upgrading is occurring with for instance, firms delivering legal services. There is overall consensus among the respondents that there is also a need to move into more complex services within different segments of services.

In fact, upgrading within the field of BPO has already taken place, especially since transactional services are now mostly automated. For instance, providers in customer services have started to engage in more complex and non-scripted activities, allowing for more judgement and adopting more complex technologies, such as smartphone application development and deployment (BPeSA, 2015; Everest Group and Letsema Consulting, 2008).

South Africa's value proposition

South Africa has been selected by client firms, predominantly from Europe and Australia, as an alternative destination for their activities. While not having the scale of India and the Philippines in terms of firms' size, South Africa is one of the few destinations offering English speaking skills at relatively low cost, and having close cultural affinity to Western markets. Furthermore, the education system as well as products and regulation in South Africa's domestic services sector are very much in line with those of its client markets, particularly the UK (Everest Group and Letsema Consulting, 2008). For instance, legal services are provided from South Africa, as the country is argued to be strong in auditing and reporting standards, and has data protection laws and certifications, similar to the UK (Willcocks *et al.*, 2015).

The cost of doing business in South Africa has dropped significantly in recent years to approximately 54 per cent cost savings for firms in the UK and Western Europe, mainly due to the incentive scheme provided by the Department of Trade and Investments (Willcocks *et al.*, 2015). However, South Africa is not trying to compete as a low cost location for large volume contracts. South Africa is aiming for a position as a destination for 'high-quality, medium-cost contact centres and back-office services' (Willcocks *et al.*, 2015). The respondents confirmed that global investors turn to South Africa for the quality of customer services, which is argued to anchor in high quality of English language skills, neutral accent and cultural affinity to suppliers' markets. As one respondent from a UK based supplier of legal services explained:

> We have work delivered by Indian agents and South Africans. Some of the bulk goes to India, but when it comes to customer interaction we like to have South African agents doing the work. […] We offshore large volume back-office work to India, but for the customer experience we choose South Africa.
> (interview with a UK based supplier of legal services)

In addition, foreign firms opt for South Africans to deliver their complex services involving judgement calls, because of their ability to undertake non-scripted work. The CEO of a customer solutions company argues that this distinguishes South Africa from more traditional offshore destinations such as India.

> We service our company out of India. When the tasks are well scripted and frameworks are in place then India is a good option, but if it deviates Indian

> employees are off. We have experience in both countries, but for our company we see more value in South Africa, because the employees are more capable to work off the script.
>
> (interview with a foreign supplier of customer solution services)

In addition, foreign service providers also chose South Africa as a delivery base. For instance, Indian service providers have opened offices in South Africa, because it offers a pool of multi-lingual skills at close proximity to the Middle East and European market (KPMG, 2011). Also, its domain skills in service delivery in financial services, particularly in actuarial science (risk management) act as a pull factor. For instance, South Africa's number of qualified actuaries is argued to be three times higher than India and the Philippines combined (BPeSA, 2015). In addition to the skills base, the growing demand for ITES in the South African and regional market itself also acts as a pull factor for foreign service providers, which will be further discussed in the next section.

Despite South Africa's capabilities in the ITES industry, particularly in customer services, some challenges need to be addressed in order to sustain the recent growth rates of the industry. Although there is a massive labour pool with generic skills, there is a shortage of experienced and adequate team leaders and senior managers. Initiatives to address this include training programmes, and individual firm strategies, such as having overseas managers on the job. In addition, firms encounter problems when it comes to retaining staff as the growing demand for employees enables them to easily switch between employers. Another severe constraint to growth of the industry is the lack of coordination and alignment across industries, which also evidences in scattered initiatives related to workforce development. For instance, respondents stress the need for cooperation between universities and the private sector, to cater to the needs of the services industry by increasing the employability of people and hence increasing the scalability of projects.

The emergence of regional value chains of services and the strategic position of South Africa

Foreign investments in the South African and SSA market

South Africa serves not only as a platform for service delivery to firms overseas but foreign service providers have also opened up offices in South Africa to access the untapped market potential in the country and the SSA region. This market potential is reflected in increasing foreign direct investments (FDI) into Africa, with services outperforming manufacturing and primary extractive industries. For instance, in 2012, services accounted for 48 per cent of the total FDI stock in the African continent, more than twice the share of manufacturing (21 per cent), mostly flowing to South Africa and Nigeria (UNCTAD, 2015). Future potential for investments has been identified in consumer services such as: technology, media, telecommunications, financial services, consumer products and retail (EY, 2015). Respondents claim that foreign interest from both service providers and client firms in the South African and regional market for services is a result of saturated markets in Western countries and a rising middle class and growing consumerism in African markets.

> First a cost saving option, but now BPO is also the result of market seeking behaviour. In Europe, there is no potential market. South Africa is an easy entry and safe option into Africa.
>
> (interview with a representative of an organization targeting skills development in the South African BPO sector)

For example, Indian service providers have explored new emerging markets in the Middle East and Africa to reduce their overreliance on customers in the United States. Consequently, Indian service providers have established a presence in the South African market to service the domestic market for outsourcing (KPMG, 2011; The Economic Times, 2010).

Foreign service providers have also expressed their interest in the South African market through acquisitions of South African firms. An example of the latter includes the UK firm Quindell, which acquired South African-based telecom outsourcing service provider Quintica (in 2012). Indian outsourcing group WNS announced in 2012 that it had acquired South African BPO provider Fusion Outsourcing Services, supporting clients both domestically and offshore (Southafrica.info, 2012b). These examples of foreign investments in the domestic market for IT-enabled services are an illustration of the significance of local value chains.

Nearshoring in SSA

Foreign acquisitions of South African firms are also driven by the interest in the SSA region. For instance, UK outsourcing service provider Capita expressed its desire to enter the SSA market starting with the acquisition of the South African firm Full Circle. In a press release the CEO of Capita explained that:

> The acquisition of Full Circle, a leading contact centre solutions business in South Africa, will enable us to provide a full range of offshored services in the region to our clients.
>
> (Southafrica.info, 2012a)

Furthermore, a study by Everest Group and Letsema Consulting (2008) claims that businesses providing outsourcing solutions in the financial services sector in South Africa have expanded their operations more rapidly into SSA, servicing firms in the neighbouring countries. This is also notable in the IT industry, where South Africa has for a long time been the major supplier of IT products and services to neighbouring countries. As such, market seeking behaviour of firms is not restricted to South Africa as a potential market, but extends to other African countries. This has often taken the form of nearshoring in SSA, referring to the outsourcing of services from firms in the region to service providers in South Africa, which give rise to RVCs. Even though the cases appear to be incidental and not well documented, respondents highlight that a lot of potential for nearshoring is present. A respondent from a South African owned IT firm in South Africa explains:

> It wasn't intentional. Our client in South Africa expanded its business to neighbouring countries, and because of that we are now also servicing our client and new clients in the African region. [...] BPO rides on the back of it.
> (interview with a South African owned ITO firm)

There are also cases where service providers in South Africa service clients in the SSA market through joint ventures with local partners in the region. For instance, Direct Channel Holdings (DCH), one of South Africa's largest privately owned BPO service providers partnered with service provider Simba Technology Ltd to deliver services in Kenya, Tanzania and Uganda. DCH's operations (call and contact centre outsourcing solutions to banking, insurance, telecoms and retail customers) are currently also provided in Nigeria through a joint venture with Nigeria's local service provider XL Management Services Group (ITnewsAfrica, 2010).

The market seeking behaviour of international service providers on the African market is not restricted to the case of South Africa, but also applies to the case of Ghana. A number of overseas foreign service providers have started operations in Ghana to service the local or regional market rather than the offshore market (for example TechMahindra, TeleTech). The foreign firms perceive the growing demand for services in Ghana and the African market, especially in financial services and the telecommunications industry, as an opportunity to increase their client base (Beerepoot and Keijser, 2015).

Shared services centres in South Africa

In addition to nearshoring in SSA, RVCs of services also consist of large companies in SSA having multiple offices across the region, which choose not to deliver their services in a dispersed matter, but bundle and perform these out of South Africa. This type of service delivery, referred to as shared services, is a form of outsourcing whereby various business functions and processes (for example in Finance and Accounting, IT, and Human Resources) from several departments, often across countries, are grouped together and provided from one location. This is generally done to achieve cost savings and business efficiencies (BPeSA, 2015). There are approximately 40 shared services centres operating in South Africa, of which 87.5 per cent are run as captive operations, and the rest are run by third-party providers (Zille, 2015). An example includes the launch of Barclays shared services centre in Pretoria, which provides operation support and services to its offices across SSA. The centre has created 4,000 jobs in voice and back-office services. Opportunities and trends have also been signalled for shared services in the field of LPO.

> Law firms have global relations, and with South Africa as a gateway to Africa the stage is set to utilize those relations.
>
> (interview with a South African legal services consultant)

> It's a matter of time before law firms move into Africa and execute their 'African strategies'.
>
> (interview with a South African owned LPO firm)

Shared services is different from nearshoring in that a centre could also perform bundled services from within a country, and that services are generally not outsourced to a third-party provider. Nevertheless, the market for shared services in South Africa is gaining momentum and appraised for demonstrating similar opportunities and capabilities as the outsourcing/offshoring industry. In some cases, a third-party provider first runs the operation before it is turned into a shared services centre. Furthermore, the networks between different branches and actors across the continent that underpin these centres are emphasised by respondents as increasing the likelihood of knowledge sharing, learning and hence capacity building in South Africa. As such, shared services are seen as integral and essential to the ITES industry.

The examples on nearshoring and shared services include both foreign firms and local owned firms in South Africa providing services in the SSA market to foreign and African client firms (often MNCs). Although limited data is available, there appears to be no clear pattern of RVCs being restricted to (African) trade blocs such as the Southern African Development Community (SADC).

RVCs and their potential as stepping stones to GVCs

The growing services industry and trade in the African region is expected to open up more opportunities for the South African ITES industry. Particularly more potential is identified in the ITO segment, as industries such as banking and telecoms need more advanced IT infrastructure management. The growing services industry in the region is also coupled with increased demand for customer relationship management and back-office services (Everest Group and Letsema Consulting, 2008). Within the African continent South Africa has the competitive edge, as it is has the largest BPO sector on the continent, the most reliable infrastructure, largest talent pool, and a more stable political and financial climate compared to other African nations (Everest Group and Letsema Consulting, 2008). Additionally, the respondents claim that South Africa, especially Cape Town, offers a 'first world experience' to its investors. Direct flights from Europe to Cape Town, luxurious expat amenities, and high-quality infrastructure set the city apart from many other African cities.

Most respondents confirm South Africa's strategic position and emphasise the pressing need to look into the African market more carefully to diversify risks, reduce overreliance on the overseas market, and secure future growth.

> The African market could be the new offshore market, which we need because globally there is a lot of competition, prices are going up in South Africa [...] and it is a way to expand our talent pool.
>
> (interview with a representative of an organization targeting skills development in the South African BPO sector)

The regional market, and especially the domestic market are argued to be important for South African owned firms, particularly small to medium sized firms (SMEs). There are successful South African owned firms, but they are mostly restricted in firm size and service mainly the domestic market. The main explanation given by respondents is the lack of resources and experience and inability to compete with large (foreign) MNCs. Furthermore, offshore deals often involve large-scale contracts whereas firms having no prior experience with large-scale (international) business processes experience difficulties in securing funds from banks.

> There are few success stories, but the African success stories on the offshore market often have a larger firm behind them to back them up in terms of resources.
>
> (interview with a representative of an industry body)

Even though the regional market appears to be not fully explored yet, the regional and especially the domestic market have already served as an entry into the global offshore industry for some local owned firms. For instance, the CEO of a South African owned firm, which employs approximately 20 people delivering IT outsourcing solutions, explained that the local and regional industry provided its management team the toolset needed to enter the global market. The provider also faced fewer difficulties obtaining a contract in the regional market compared to the global market, because its contract with the client firm in the region was the result of prior established relationships in the domestic market.

Respondents from UK owned service providers located in South Africa argued that for them the African market is not used as a stepping stone into the global market. They perceive the regional market more as risk diversification and an opportunity to expand their global services delivery capacity. In fact, some highlighted the need for local expertise to expand in Africa. A respondent from a foreign firm in South Africa, servicing the offshore and local market, explained how they were looking to expand operations into the rest of Africa servicing neighbouring countries, but needed resources and local expertise to do so.

> Knowledge is needed, in particular local expertise to roll out to the rest of Africa. [...] Partnership is important, learning from other operations, what they know.
>
> (interview with a contact centre subsidiary of a UK firm)

Several foreign service providers in South Africa have been cooperating closely with South African firms to share best practices, marketing leads and business developments to expand the local and regional industry (BPeSA, 2012). Furthermore, foreign investors setting up a captive unit have often done this with the help of local third-party providers, showing that South African firms may indeed have a better understanding of the domestic and regional market.

Discussion: implications for outsourcing and offshoring

The case of South Africa shows that firms have offshored their services operations to South Africa either through a captive unit or through an external service provider. In addition, foreign service providers from advanced and emerging economies increasingly use South Africa as a base not only to service the overseas market, but also the South African market, giving rise to local value chains of IT-enabled services. These LVCs, in turn, have demonstrated the potential to be extended to RVCs as service providers seek to expand their delivery capacity further into the African region. This occurs particularly through nearshoring and through shared services centres.

The case of South Africa further shows that it is not solely large MNCs in advanced markets offshoring to developing countries to achieve business efficiencies. Rather, the case demonstrates the importance of the South African and SSA market for IT-enabled services as well. There is thus a need for a more elaborate overview of outsourcing/offshoring modes in which different markets (domestic, regional and global) come to the fore and in some cases overlap. Table 10.4 provides an overview of different sourcing modes when taking into account the heterogeneity of value chains. In each mode, the type of provider, viewed from the perspective of South African based service providers, could be an independent provider or a captive unit. The location of the client depicts the end market of the service, hence, giving rise to different value chains (local, regional and global value chains).

Although South Africa's share in the global offshore market has grown in recent years, its engagement in LVCs of IT-enabled services is currently more significant than its engagement in GVCs. Furthermore, RVCs for IT-enabled services are emerging and gaining attention, illustrating that South Africa does not solely act as an export platform for services to advanced nations. This shows the growing importance of developing and emerging country markets in the outsourcing/offshoring industry, arguing that this change in geographies of services is not well reflected in the literature.

Conclusion

The case of South Africa is breaking with the mainstream understanding of the current wave of outsourcing/offshoring focusing mainly on GVCs of services (see Massini and Miozzo, 2012; Gereffi *et al.*, 2010; Oshri *et al.*, 2015; Lambregts *et al.*, 2015), which involve the relocation of services from firms in advanced countries to less developed countries. Instead, local and regional value chains of

Table 10.4 Service providers in South Africa: heterogeneous value chains

		Location of client		
		Domestic market	*Regional African market*	*Global market*
Type of provider	Local or foreign independent (third-party) provider	*Onshore outsourcing services* South African-based firm provides services to client firm in South Africa	*Regional outsourcing services (nearshoring)* South African-based firm provides services to client firm located in SSA, shared services for multiple units of the client firm	*Global outsourcing services* South African-based firm provides services to overseas client firm (outside of SSA)
	Captive unit of a local or foreign firm	*Onshore captive services* Captive unit in South Africa provides services to the parent company in South Africa including onshore shared services centres	*Regional shared services* Captive unit of a firm in South Africa provides bundled services to other affiliates of the firm across SSA countries	*Offshore captive services* Captive unit of a foreign client firm provides services to the parent firm overseas

IT-enabled services are equally important for firms based in less developed countries. The domestic and regional ITES market allow some firms to build capacity and gain experience to compete in the global market place. This finding is line with advocates of RVCs who argue that RVCs serve as a stepping stone into GVCs (see Bamber *et al.*, 2013; Cattaneo *et al.*, 2010; ERA, 2009; Kaplinsky and Farooki, 2010). However, the examples documented in this research show that client firms in the regional market are both African firms as well as foreign firms, predominantly large MNCs. This raises questions as to how different RVCs are from GVCs in terms of governance structures (such as the power relations between client firm and provider). Additional and more detailed research is needed that includes a large number of cases to explore what RVCs look like in terms of types of service providers and client firms and the required skills and experience, and compare these to the characteristics of the global offshore market (GVCs). In other words, to what extent RVCs are different from GVCs of services and whether they indeed offer more opportunity for African operations to compete in the outsourcing and offshoring industry. Questions also arise concerning the sources of knowledge and innovation underpinning firm competitiveness. For instance, what role these different value chains play and the different learning opportunities resulting from these different networks between provider and client firm. This subject is still understudied while it is crucial for understanding firm competitiveness in the current economic landscape where learning and innovation are key (see Ernst, 2002; Morrison *et al.*, 2008; Pietrobelli and Rabelloti, 2011).

The ITES services industry in South Africa is relatively small, but has expanded rapidly in recent years. South Africa has the potential to attract more (international) investments in the IT-enabled services industry provided that it can successfully position itself to deliver higher value services. Yet, integral to this and to upgrading trajectories is human capital development. Challenges related to workforce development (for example management training and skills development) need to be tackled, as well as the lack of coordination and alignment between stakeholders across industries. Furthermore, the South African ITES industry needs to be better marketed not just in terms of South Africa's capabilities in the field of service delivery, particularly in customer services and its domain skills in the financial industry, but also in the SSA market for nearshoring and shared services. South Africa acting as a gateway to the African market could allow firms to pursue their 'Africa strategies'. The latter has the potential of becoming the country's major strengths and unique value, within the changing global sourcing landscape where MNCs seek to expand their global delivery capacity (see Gereffi *et al.*, 2010; Sako, 2009).

Note

1 Domestic captives refer to 100 per cent subsidiaries of firms in South Africa, which provide services exclusively to the parent company. Offshore captives refer to 100 per cent subsidiaries of companies based outside South Africa, which provide services exclusively to the parent company outside South Africa.

References

Adelakun, O. and Iyamu, T. (2013). Offshore IT Outsourcing to South Africa Analysis of Readiness and Attractiveness, International Conference on ICT for Africa, Harare, Zimbabwe.

Africa report (2012). Outsourcing: the Indian connection and Africa. Retrieved from www.theafricareport.com/North-Africa/outsourcing-the-indian-connection-and-africa.html [accessed 21 October 2015].

Bamber, P., Fernandez-Stark, K., Gereffi, G. and Guinn, A. (2013). Connecting Local Producers in Developing Countries to Regional and Global Value Chains: Update, *OECD Trade Policy Papers, No. 160,* OECD Publishing.

Beerepoot, N. and Keijser, C. (2015). The service outsourcing sector as driver of development: The expectations of Ghana's ICT for accelerated development programme. *Tijdschrift voor economische en sociale geografie,* 106(5), pp. 556–569.

BPAP (2015). IT-BPO Road Map 2011–2016. Retrieved from www.ibpap.org/about-us/it-bpo-road-map-2011–2016 [accessed 20 October 2015].

BPeSA (2012). BPO giant Aegis takes Cape Town SMME under its wing. Retrieved from www.bpesa.org.za/2012/07/13/bpo-giant-aegis-takes-cape-town-smme-under-its-wing [accessed 20 October 2015].

BPeSA (2015). Key indicator report 2014/2015. BPeSA, South Africa.

Cattaneo, O., Gereffi, G., and Staritz, C. (2010). *Global Value Chains in a Postcrisis World: A Development Perspective*, Washington, DC: World Bank.

Dihel, N. (2011). Can Kenya become a global exporter of business services? International Centre for Trade and Sustainable Development, 10(9), pp. 19–21.

Dossani, R. and Kenney, M. (2007). The Next Wave of Globalization: Relocating service provision to India. *World Development,* 35(5), pp. 772–91.

ERA (2009). Developing African Agriculture Through Regional Value Chains, United Nations Economic Commission for Africa Addis Ababa, Ethiopia.

Ernst, D. (2002). Global production networks and the changing geography of innovation systems. Implications for developing countries, *Economics of Innovation and New Technology*, 11(6), pp. 497–523.

Everest Group and Letsema Consulting (2008). South Africa's BPO proposition: discussion document. London: Everest Group.

EY (2015). EY's attractiveness survey Africa 2015 making choices, Ernst and Young Global Limited.

Fernandez-Stark, K., Bamber, P. and Gereffi, G. (2011). The Offshore Services Value Chain: Upgrading Trajectories in Developing Countries. *International Journal of Technological Learning, Innovation and Development*, 4(1-2-3), pp. 206–34.

Gereffi, G. (2005). *The New Offshoring of Jobs and Global Development.* Geneva: International Labor Office.

Gereffi, G. and Fernandez-Stark, K. (2010). *The Offshore Services Global Value Chain,* Center on Globalization, Governance and Competitiveness, Duke University.

Gereffi, G., Fernandez-Stark, K. and Psilos, P. (2011). *Skills for Upgrading: Workforce Development and Global Value Chains in Developing Countries*, Center on Globalization, Governance and Competitiveness, Duke University.

Gereffi, J., Humphrey, J., Kaplinsky, R. and Sturgeon, T. (2001). Introduction: Globalisation, Value Chains and Development, *IDS Bulletin,* 32(3), pp. 1–9.

ITnewsAfrica (2010). SA IT company invests in Nigeria. Retrieved from: www.itnewsafrica.com/2010/05/sa-technology-company-invests-in-nigeria [accessed 21 October 2015].

Harji, K., Best, H. and Jackson, E-T. (2013). *Digital Jobs*, The Rockefeller Foundation.

Humprey, J. and Schmitz, H. (2002). How does the insertion in global value chains affect upgrading in industrial clusters? *Regional Studies*, 36(9), pp. 1017–27.

Kaplinsky, R. and Farooki, M. (2010). Global Value Chains, the Crisis and the Shift of Markets from North to South. In: Cattaneo, O., Gereffi, G., and Staritz, C. (2010), *Global Value Chains in a Postcrisis World: A Development Perspective* (pp. 125–154). Washington, DC: World Bank.

KPMG (2011). India-South Africa Corridor Newsletter, Building Relations, KPMG International Cooperative.

Labour Research Service (2011). South African MNCs in Africa – Trends in 2010 / 2011: A report covering MNCs submitted by Global Unions. Cape Town: Labour Research Service.

Lambregts, B., Beerepoot, N. and Kloosterman, R.C. (eds) (2015). *The Local Impact of Globalization in South and Southeast Asia: Offshore Business Processes in Services Industries*. London and New York: Routledge.

Mann, L. and Graham, M. (2016). The Domestic Turn: Business Processing Outsourcing and the Growing Automation of Kenyan Organisations. *Journal of Development Studies*, 52(4), pp. 530–48.

Massini, S. and Miozzo, M. (2012). Outsourcing and Offshoring of Business Services: Challenges to Theory, Management and Geography of Innovation, *Regional Studies*, 46(9), pp. 1219–42.

Morrison, A, Pietrobelli, C. and Rabellotti, R. (2008). Global Value Chains and Technological Capabilities: A Framework to Study Learning and Innovation in Developing Countries, *Oxford Development Studies*, 36(1), pp. 39–58.

NASSCOM (2015). India IT-BPM Overview. Retrieved from www.nasscom.in/indian-itbpo-industry [accessed 20 October 2015].

OECD (2015). The Participation of Developing Countries in Global Value Chains: Implications for Trade and Trade Policy, Paris: Organisation of Economic Cooperation and Development.

Oshri, I., Kotlarsky, J., and Willcocks, L.P. (2015). *The Handbook of Global Outsourcing and Offshoring*, 3rd Edition, Palgrave Macmillan.

Palugod, N. and Palugod, P. (2011). Global trends in Offshoring and Outsourcing, *International Journal of Business and Social Sciences*, 2(16), pp. 13–19.

Pietrobelli, C. and Rabelloti, R. (2011). Global Value Chains Meet Innovation Systems: Are There Learning Opporunities for Developing Countries? *World Development*, 39(7), pp. 1261–69.

Ramdoo, I. (2014). Developing value chains: What role for regional integration? *GREAT Insights*, 3(7).

Sako, M. (2009). Global Strategies in the Legal Services Marketplace: Institutional Impacts on Value Chain Dynamics, Paper presented at the SASE Conference. Paris, France. 16–18 July.

Southafrica.info (2012a). UK outsourcing giant buys into SA. Retrieved from www.southafrica.info/business/investing/capita-250712.htm#.Veax8nhUPNo#ixzz3kZGHyWSm [accessed 19 October 2015].

Southafrica.info (2012b). SA's outsourcing reputation grows. Retrieved from www.southafrica.info/business/investing/opportunities/outsourcing-260912.htm#.VebCgnhUPNo [accessed 19 October 2015].

Sturgeon, T. and Memedovic, O. (2010). Measuring global value chains: intermediate goods trade, structural change and compressed development. Vienna: UNIDO working paper.

The Economic Times (2010). Indian IT companies like TCS, Infosys and Wipro tap Africa's new-fangled fad for technology, Retrieved from http://articles.economictimes.indiatimes.com/2014-05-10/news/49757824_1_hcl-technologies-south-africa-finacle.

UNCTAD (2015). World Investment Report 2015: Reforming International Investment Governance, United Nations Conference on Trade and Development, Geneva.

Wausi, A., Mgendi, R. and Ngwenyi, R. (2013). The labour market and digital jobs in Africa: Kenya's workforce potential in Impact Sourcing, *Research Brief June*, No.3.

Willcocks, L., Lacity, M. and Craig, A. (2015). *South Africa's BPO Service Advantage*, Palgrave London.

Zille, P. (2015). Analysis of the Business Process Services Sector, *Genesis Analytics*, Johannesburg.

11 Services offshoring and skill development in emerging economies

Evidence from Metro Manila, Philippines

Mary Leian Marasigan and Bart Lambregts

Introduction

Services offshoring, or the cross-border translocation of services production, notably from developed to developing countries, has generated several millions of export-oriented services jobs in the latter.[1] The IT-enabled trend, which kicked off in earnest in the 1990s and gained momentum after the turn of the century, has left few countries unaffected (Oshri *et al.*, 2011; see also Tholons 2014). The work that is shifted from high-wage to lower-wage economies is multifarious and, apart from industry-specific activities, tends to fall into the generic domains of information technology outsourcing (ITO), business process outsourcing (BPO) and knowledge process outsourcing (KPO) (Fernandez-Stark *et al.*, 2011). It typically concerns so-called 'back-office' work, i.e. work that requires little or no face-to-face contact with clients, suppliers or even co-workers and the fruits of which can be electronically transferred. While this kind of work in advanced economies is generally catalogued as 'low-skilled', 'low-value-added' and/or 'routine', for developing countries it often represents work that previously was performed either not at all, on a modest scale only, or at lower service level standards (Fernandez-Stark *et al.*, 2011).

Entry in global production networks and the concomitant arrival of new kinds of work not only generate employment, but – at least potentially – also create opportunities for the upgrading of existing and development of new skills by members of the labour force. Especially when these skills are portable, i.e. usable beyond the job context in which they are acquired, the gains, both for the individuals involved and the economy at large, can be substantial and more durable than the initial employment effect itself (Bottini *et al.*, 2007). Diversification and enhancement of skills in a country's labour force are likely to lead to productivity gains and increased competitiveness, and enable entry into new types of production and upward movement in global value chains. If, in addition, parts of the monetary gains are used to provide and pursue better education, the development benefits may accrue to even more (ILO, 2014). Yet, such favourable outcomes are by no means guaranteed. If much of the new work is routine and low-skilled, and the employed skills themselves of little use elsewhere in the economy, the more durable employability gains identified above are not likely to materialise. The picture becomes bleaker still if, attracted by comparatively good remuneration, such routine jobs are performed by what in theory should be the country's brightest (i.e. those having completed tertiary education). In that case, the impact on development could even be negative, with talent being squandered in low-value-added work and not being available for – from a development perspective – more constructive kinds of work (Bird and Ernst, 2009; Bottini *et al.*, 2007; Hayter, 2004).

While there exists an extensive literature about how human capital matters for offshore services production (see for example Chadee *et al.*, 2011; Fernandez-Stark *et al.*, 2010; Graf and Mudambi, 2005, Hollinshead *et al.*, 2011), the question how, on the other hand, services offshoring affects human capital development in the receiving countries has only recently emerged as a topic of scholarly interest. Preliminary answers given so far vary, with some arguing that due to its elementary and routine nature work in offshore services production hardly adds to employees' skills portfolios (e.g. Bird and Ernst, 2009; Taylor and Bain, 2005) and others such as Beerepoot and Hendriks (2013), Marasigan (2015) and Vira and James (2012) finding that people working in the offshore services industry do accumulate new skills and manage to enhance their employability. The empirical base underlying these seemingly opposing perspectives, however, is still rather thin meaning that additional empirical studies are in great demand (Beerepoot *et al.*, 2015; Bottini *et al.*, 2007; Fernandez-Stark *et al.*, 2010).

This chapter empirically engages with the question whether and how services offshoring contributes to human capital development in developing countries. We focus on one of the world's largest beneficiaries (in terms of employment gains) of the services offshoring trend, namely the Philippines, and use interview and survey data collected among dozens of BPO workers in Metro Manila to examine: a) how the industry utilises the skills of its workers; and b) if and how workers' skill portfolios are broadened and/or enhanced through working in the industry. The answers will help to advance the debate about labour and skills related externalities generated by services offshoring.

The chapter from here breaks down in five parts. In the next section we briefly summarise the literature on labour skills effects generated by services offshoring. We subsequently explain our research methods. We then use three sections to present our empirical findings before drawing conclusions.

Services offshoring and workforce skill development

The impacts associated with a local economy becoming integrated in global production networks are many. They include the creation of direct and possibly indirect employment, the establishment of forward and backward linkages with other local firms, technology and knowledge spill overs (diffused via said linkages and via labour mobility), wage effects, labour productivity effects (global integration often bringing new types of work which frequently are characterised by higher labour productivity than much of the already available work), and eventually – and hopefully – local, regional and perhaps even national welfare effects (Bottini *et al.,* 2007). Most certainly part of the package are human capital development effects, or in other words: the impact insertion into the global economy may have on the productive capabilities of the local workforce.

With human capital being the key input in services production, local workforce skill levels determine to a large degree the nature and the complexity of services tasks that can be offshored to a specific destination. The quality of the local workforce, in other words, determines at which level of value-added a local economy may secure a position in global value chains or production networks (Chadee *et al.*, 2011; Fernandez-Stark *et al.*, 2011; Graf and Mudambi, 2005; Hollinshead *et al.*, 2011). In cases where services tasks are offshored to developing countries, insertion into global value chains usually happens at the low-skilled end (Bottini *et al.*, 2007). For the receiving economies the work involved, however, may be (partly) new or, if not new, require a higher level of workforce competence in its production than similar services produced for the domestic market (ibid.). This means that even entry at the lower end of global services value chains for local economies and their workforce may mark a first step on a trajectory towards workforce skills upgrading and human capital development.

People acquire productive capabilities in various ways, including through formal learning in education systems, socialisation (e.g. learning from friends or family), formal learning in the workplace (e.g. training programmes) and learning by doing (ILO, 2014). In low-income countries, the average level of educational qualifications has risen substantially over the past few decades (Barro and Lee, 2010). In addition, in some countries responsive governments and educational institutes, often encouraged by industry actors, have started to attune educational programmes to the (skill) needs of leading industries (Kleibert, 2015). While these developments facilitate entry in global services production networks, often a gap remains between the skill levels of prospective workers and the levels required by the services offshoring industry itself. As noted, the international clients serviced by offshore services providers often demand higher quality standards than is common in developing countries' domestic markets. Offshore services firms respond by recruiting workers with a minimum level of education and competence, and by investing in additional workforce development measures (Fernandez-Stark *et al.*, 2011). Such measures may include basic induction sessions preparing new employees for the tasks they will perform, but also more advanced (on-the-job or externalised) training programmes aimed at acquiring technical or managerial skills, help in skill certification, mentoring and providing access to advanced formal degree programmes such as MBAs or other Master's programmes (ibid.; Wadhwa *et al.*, 2008).

Skill development initiatives for workers may differ depending on the offshore services segment a firm operates in (for example BPO, ITO or KPO) and on the firm's position in the value chain. As shown by Bird and Ernst (2009), employee training and development in IT-BPO and BPO differs substantially. In call centres, for instance, new employees receive introductory training consisting of generic English proficiency training, basic customer service skills, and possibly an introduction to the culture and geography of the market that the employee is going to service. Furthermore the employee will receive product-specific training, focused on the product he or she will be responsible for. Periodical updates may be received too, depending on the nature of the product. On top of these, (selected) employees may receive management training, if they are seen to have leadership potential. Training may be produced in-house, or sourced from external training providers. Alternatively, employee training in ITO firms is heavy on technical (IT) skill development, with frequent updates and, if the firm operates in a higher value-added segment of the market, a focus on the acquisition of up-to-date certifications and training that enhances problem-solving skills and innovative thinking (Fernandez-Stark *et al.*, 2011).

The central idea is that the 'on-the-job' development of additional capabilities enhances labour productivity and increases opportunities for firms and the industry at large to diversify into new services tasks, apply more advanced technologies and production processes, and move up into higher value-added parts of the value chain. Success would then trigger yet other rounds of investment in human capital development, ideally not only by firms but also by the educational sector (investing in higher quality formal education in an effort to keep up with market demand) and by (prospective) workforce members and their families (investing in education at the household level). Domestic entrepreneurship, in the process, may be stimulated as well (Bottini *et al.*, 2007; ECLAC, 2008; Fernandez-Stark *et al.*, 2011).

However, when the market environment lacks opportunities (e.g. when there is little scope for either upgrading in the value chain or horizontal expansion in adjacent markets) or when local actors fail to take advantage of the opportunities presenting themselves (e.g. by not investing surplus income in productivity enhancing strategies), less favourable outcomes are possible as well. A local economy and workforce specialising and becoming locked in specific low-value-added, low-skilled and labour-intensive tasks could impede the development of more advanced skills and hence increase the skills gap with other economies (Bottini *et al.*, 2007; Hayter, 2004). In today's services offshoring practice, where the offshored tasks tend to be narrowly defined rather than composite, this is more than just an imaginary threat. If, moreover, in such cases available local human capital is misallocated or underutilised, for instance as a result of such work paying relatively well and attracting workers that are essentially overqualified, the human capital development impacts could become more adverse still.

The impacts of the more favourable upgrading and development scenario can be further strengthened and the effects of the less favourable 'lock-in' scenario softened if the skills acquired by service offshoring workers are portable or transferable (i.e. skills that are usable beyond the job context where they were acquired, such as (English) language skills, analytical skills, problem-solving skills, communication skills). From a broader human and economic development perspective, so it can be argued, the skills acquired by workforce members being portable matters as much, perhaps even more, as their utility in enhancing productivity in the services offshoring industry itself. Because of its relative low capital intensity and its relatively limited dependence on local suppliers, the offshore services industry in theory is rather footloose (Bottini *et al.*, 2007; UNCTAD, 2004). The discovery of greener pastures elsewhere or a reversal of the services offshoring trend may put a quick end to a local offshore services boom. Portable skills (in contrast to skills only useful in a particular offshore services context) will remain useful also when the jobs in which they were acquired have disappeared. As such they make for important potential externalities that may enhance the employability of the workers involved (Marasigan, 2015) and result in the upgrading of an economy's workforce at large (Fernandez-Stark *et al.*, 2011).

Below, we empirically examine how the BPO industry in Metro Manila utilises the skills of its workers and affects workers' skill portfolios.

Methodology

Data for this study was collected by way of interviews and a survey, with Manila-based call centre and back-office support workers being the respondents (call centre work and back-office support belong to the BPO domain). The interviews took place from January till March 2013 and from June till October 2014. Purposive sampling by means of snowballing was used, starting with the contacts of the first author. Since current workers may view their working experience differently than former workers the interview sample was made to include both categories (28 currently working and 32 former BPO employees). The interviews were held in coffee shops outside of the company, in fast food joints, or in workers' residence. All interviews lasted between 30 and 45 minutes and, except for one, were audio recorded and transcribed in full.

The survey was run after the first round of interviews, between January and June 2014, and targeted currently employed workers. It included 92 questions on workers' profile, educational background, current job, and employment history. The questionnaire was pre-tested online and revised based on the results and feedback from respondents. Hard copies of the questionnaire were then distributed to BPO workers who were at the same time masters students at the University of the Philippines School of Labor and Industrial Relations (UP SOLAIR), and to HR officers from four BPO firms who volunteered to enlist the participation of BPO colleagues, staff and friends in the survey. Eventually, 394 questionnaires from more than 50 different BPO companies were returned, from which 358 forms were usable. Finally, five BPO management personnel (four from HR and one from Business Development) were interviewed to obtain a firm-level perspective on BPO work. These interviews were also fully transcribed.

A profile of currently employed BPO workers and their skills

The workers who participated in the survey are overwhelmingly young, single, and highly educated. Of the total, 70 per cent are less than 30 years old, 69 per cent are single, and 56 per cent are female. Forty-three per cent work in call centres, 56 per cent in back-office support, and 1 per cent in software development. As for job level, 77.7 per cent hold entry-level BPO positions with titles such as customer service agent/representative, business analyst, technical support representative, accounting specialist, HR associate, sourcing analyst, etc. Sixty-four per cent of respondents consider the current job as a possible long-term job or career, 20 per cent think of it as a short-term job, and 13 per cent view it as a training ground. The current job is the first work experience of 21 per cent of respondents while 77 per cent have held a previous job (2 per cent non-response).

Table 11.1 shows that the number of respondents who had no other work experience aside from the BPO almost equals those who had some experience in other sectors before joining the BPO. Around 20 per cent of respondents has benefitted from pre-BPO training provided by the government, most of which (76 per cent) concerned 1–2 months of call centre training. This indicates that workers, in addition to their educational qualifications have had previous opportunities to learn other skills that may or may not have a relation to their current job.

BPO workers furthermore have high levels of educational attainment: 99 per cent of total respondents attended at least a few years of college, 84 per cent completed college, and 12 per cent pursued post-graduate degrees (quite similar as observed by Bird and Ernst (2009) and Magtibay-Ramos *et al.* (2008)). A closer examination of college programmes that workers attended reveals that the majority studied Business Administration and related courses (24.6 per cent), Social and Behavioural Science (20.7 per cent) or Information Technology Education (19.0 per cent). Others reported to have studied Medical and allied courses (10.9 per cent), Mass Communication and Documentation (6.7 per cent), Engineering and Technology (6.7 per cent), or Education and Teacher Training (3.4 per cent). This distribution of respondents based on fields of study resembles the national trend in terms of student enrolment in the Philippines.

With the large majority of BPO workers having completed college or even higher levels of education, and with entry-level call centre and back-office support positions only requiring a high school diploma to two years of college (DOLE,

Table 11.1 Previous work experience of survey respondents (n=358)

Work experience	*%*
Has had a BPO job since leaving school	21
In and out of employment but all jobs were in the BPO	12
BPO, non-BPO, and currently BPO	6
Non-BPO* and currently BPO	34
Worked abroad before joining BPO	1
No answer	25
Total	100

Note: *Top non-BPO sectors respondents (n=172) worked in are: restaurants, hotel, travel, tourism (15%); banking, finance, insurance (12%); construction, mining, manufacturing (12%); retail and distribution (12%); health and medical services (10%); and education, training, academe (9%).

2012), the suggestion is that many BPO workers in terms of formal educational attainment are overqualified for the entry-level BPO work that they perform. Yet, simultaneously, the rate of immediate hiring (that is the share of applicants judged directly employable by BPO firms looking for staff) is only 5 per cent (Hernandez, 2012). Applicants bringing along insufficient skills is often quoted as the main reason (Visconti, 2012). This apparent paradox, of (prospective) BPO workers in the Philippines 'on paper' being overqualified but in practice often being judged underqualified for the work they (wish to) do, merits a closer look at the skills people are supposed to acquire through higher education in the country.

In the Philippines, the Commission on Higher Education (CHED) mandates the policies and guidelines including curriculum content, faculty qualifications, and graduates' competency standards for each college course offered by public and privately-owned higher education institutions (HEIs) in the country. The CHED therefore, defines (in part) the baseline knowledge and skills that HEIs are expected to deliver to the labour market. Analysis of the top six disciplines that respondents majored in reveals a great variety of competency standards in terms of content and form. Table 11.2 shows that this holds for the more programme-specific technical skills (which should come as no surprise) as well as for generic skills. For instance, communication skills, which are critically important in many BPO segments, are considered to be a major competency area under the BS Business Administration and BS Information and Technology Education (ITE) programmes. The latter includes the recently introduced specialisation programme called 'Service Management', designed specifically for those who wish to join the BPO sector (CHED, 2012). Meanwhile, in the BS Teacher Education communication skills are lumped together with other generic skills: 'basic and higher level literacy, communication, numeracy, critical thinking, learning skills...' (CHED, 2004, p. 2). The BA in Communication in turn, quite interestingly does not mention communications skills as a competency area at all. Another valued generic skill – problem solving – does not appear in the listings at all, although reference is made to closely connected 'judgement and decision-making' and (critical) 'thinking' skills in some fields.

Table 11.2 suggests that the college programmes concentrate on teaching technical skills and reveals that so far little thought has been given to developing a cross-disciplinary baseline set of generic skills for college graduates. With generic skills also having been identified by the World Bank as the domain where education in the Philippines underperforms most (Di Gropello, 2010), the gap perceived by BPO firms between the skills promised by a college degree and the skills needed in the workplace, should perhaps not come as a surprise.

Notwithstanding, a majority of the surveyed workers acknowledge the impact of education on work performance. To the question how useful they considered their educational background for the job they were doing, 22.1 per cent answered 'very useful', 24.3 per cent 'useful' and 23.5 per cent 'somewhat useful', with back-office support workers and call centre agents returning roughly similar responses. These numbers however also suggest that there is a substantial minority who think otherwise.

Table 11.2 Generic and technical skills developed by college programmes completed by BPO workers

Specific course	*Generic skills*	*Technical skills*
BS in business and BS in information technology education, service management specialisation track	English communication skills (oral and written); access, retrieve, disseminate information using IT; interpersonal relations; positive work ethics; thinking skills; leadership.	Conduct feasibility studies; prepare business plans; conduct environmental scanning; and the concepts, approaches, and techniques of environmental conservation.
BA in political science	Writing and presentation skills.	Understanding of major concepts in the discipline; thorough knowledge of political science, sub-disciplines, themes, techniques; conduct scholarly research; substantive understanding or the historical and contemporary local and global developments.
BA in communication		Planning, managing, and evaluating communication programmes; design/produce multimedia; design, evaluate communication campaigns; design media-based learning systems, conduct communication research.
BS in medical technology	Judgement and decision-making; data processing and encoding; communication; human relations skills; management and leadership skills.	Collect samples, specimens for analysis; determine acceptability of samples within guidelines; perform tests according to techniques; observe principles of data security or patient confidentiality.
BS in teacher education	Literacy; communication; numeracy; critical thinking; creativity; innovativeness; and learning skills.	Understanding learning processes; role of teacher in processing learning; knowledge of the subject matter; teaching process skills; facilitate learning of diverse types of learners.

Source: CHED Memorandum Orders 39-2006; 34-2012; 31-2011; 10-2010; 14-2006; 30-2004, available at www.ched.gov.ph/index.php/issuances/cmo.

Skills utilisation in BPO work

Irrespective of BPO firms' dissatisfaction with the skill levels presented by most job applicants, those eventually hired do bring along a portfolio of skills. These include the generic and technical skills acquired through formal education as well as any possible skills gained in previous jobs or through other life experiences (Inkson and Arthur, 2001; Sullivan, 1999). Which of these skills are being utilised by the BPO sector? Various authors have shown that offshore services cater to different industries and involve activities located in various places along the value chain (Batt *et al.*, 2009; Gereffi and Fernandez-Stark, 2010; James and Vira, 2010). This means that in theory it should be possible for workers to find employment that relates to their field of study and appeals to not only their generic but also their technical skills.

Our data reveals that on joining the BPO industry, workers hope that they will be able to use what they learned in college (technical skills), but they are also prepared to accept a job that is less appealing to them, especially if they want to land a job as fast as possible. Job search intensification in order to find a desired job entails additional costs and goes without guarantees that a better match will be found. Rather than persisting on finding a desired position, workers focus on other subjective benefits of the job being offered such as commuting distance from their place of residence and company 'brand'. If they are unable to obtain a desired post, first-time workers are especially likely to take the next best offer. As one applicant recalled in relation to the first job offer he received:

> All throughout the recruitment process I thought I was being considered for the HR position because I applied for that, but just before I was offered a job I was told they were only hiring procurement specialists and then they asked me if I would like to take that position. I thought, after having come this far I better just take their offer.
>
> (Male, former back-office support analyst, 12 January 2013)

The background to this is that BPO firms are often in a hurry to fill job openings and hence willing to channel 'generically qualified' applicants to positions that do not necessarily offer the best match to their 'technical' backgrounds and/or preferences. It suggests that for BPO firms, applicants' generic skills trump technical skills, and means that (prospective) BPO workers cannot expect their technical skills to be decisive in the hiring process which in turn reduces the chance that people will have an opportunity to enhance their existing technical skills in BPO work.

The job applicant screening tests are indicative of the kind of skills that are required by the BPO sector. The process usually involves several rounds of interviews. These may be over-the-phone, one-on-one, and with a panel of several interviewers. Applicants may also have to perform a typing-speed test and answer a simulated call from a mock customer. An HR manager explained that they first evaluate how an applicant sounds on the phone. Only if this assessment is favourable, the recruitment specialist looks into the educational background of the applicant (female, VP HR, 6 Feb 2014). Job applicants for back-office support positions undergo written tests for maths, English, and reasoning in addition to several interviews (female, former back-office worker, 14 Mar 2013). Back-office workers are expected to bring along slightly higher formal qualifications and skills competence than call centre agents (see also Gereffi *et al.* 2011). Summing up, the basic skills required for BPO work are communication skills and basic IT skills (being able to type), while back-office support workers need to bring acceptable levels of maths, literacy, and IT skills. Beyond skills, recruiters also evaluate applicant demeanour and willingness to do night work.

The workers' job tasks elucidate further on the kinds of skills that are utilized in BPO work. Table 11.3 presents a (representative) sample of job descriptions provided by interviewed workers. The job content reports presented suggest that the key skills required to perform the work are proper communication and/or typing skills, a good dose of accuracy and in specific cases numeracy and/or domain knowledge (e.g. the lease contracts and insurance accounts). A comparison with the skills taught in various college programmes as presented in Table 11.2 learns that the BPO jobs almost exclusively appeal to college graduates' generic skills, and in fact to quite a narrow selection of these alone. Technical skills are hardly or not being called upon. The general impression emerging from this is one of skills underutilisation: college graduates working in the BPO industry would seem to employ only a fairly small part of the skills they acquired in their formal education.

Workers' perceptions of how the job appeals to their skills seem to be influenced by the nature of the account they work for and by whether they are new on the job. Call centre agents expressed more distinct views than back-office support workers. Three respondents working in the same telephone directory assistance account mentioned that the job 'quickly dulls the mind' (16 Aug 2013). While they have to work fast – as each call must be done in under 30 seconds – the job being highly routinised does not appeal to many skills other than mastering the use of the database and memorising places in the US and Canada to not mistake the use of one database for the other. Alternatively, some new call centre agents think that their job has a very steep learning curve and calls forth different skills at the same time: communication, problem solving, IT skills, and multi-tasking (female, former call centre agent, 13 Nov 2013). Meanwhile, four seasoned call centre agents (active and former) who had worked in different accounts observed that while each account is challenging at the beginning, it becomes easier over time because 'you talk about the same (product) over and over' (female, call centre agent, 16 Feb 2013). Two current agents (both college undergraduates) mentioned that there comes a moment when they would be able to do the job almost with their 'eyes closed' adding that when work becomes '*petiks*' (child's play) some agents are triggered to transfer to a different firm.

Table 11.3 Selection of BPO job tasks, account type and workers' educational qualifications

Educational qualification	*Account type*	*Job content as described by respondents*
Elementary school teacher	Directory assistance	We ask callers questions like 'what city and state, or what listing, or what are you looking for?'
Nursing graduate	Customer service for a US department store	We place orders for customers who ask for assistance, have trouble using the website, or have received a wrong item or failed to receive their order.
Legal management graduate	Transcription	Our clients (publicly listed US firms) have quarterly calls. Our voice writers transcribe these calls. This means that agents listen and dictate what they hear to software that then produces the printed transcript.
High school teacher	Insurance	We answer calls about eligibility and benefits from plan holders.
Industrial engineer	Supply chain	I process purchase orders from foreign buyers. I check the price, delivery date and if the (seller) can deliver the materials according to the specifications of the buyer.
Mathematics graduate	Lease contracts	We prepare abstracts of lease contracts between a building owner and its tenants. This includes computing for the remaining years of the lease, the rent that must be paid, interest, and costs for overstaying.
IT graduate	Legal database	We create links between new US judicial courts decisions and past related cases.
Nursing graduate	Credit collection	We call up customers in the US to collect overdraft charges.
Electronics engineering graduate	Software testing	We test software for possible problems including presence of bugs.
College student	Online English tutorial	There are three positions in our company – mock students, tutors and performance evaluators. Mock students pretend to be Japanese students. As a performance evaluator I listen to recorded tutorials and rate tutors' performance based on several criteria.
Behavioural sciences with masters units	Manpower recruitment	I examine CVs to find qualified persons for different client firms, interview applicants, and negotiate salary rates between our client and job applicants.
Nursing undergraduate	Technical Support	Customers call us to get help to solve hardware-related computer problem like a defective monitor.
College undergraduate	Credit investigation	Our clients are banks or loan institutions. We call the companies (potential customers of our clients) and ask a few questions to see if any of them sounds bogus.
Law student	Book indexing	I once prepared the back of a book index for a UK law book.

In contrast, workers in back-office support perceive that their job is not any different from typical office-based jobs although they work nights. Like other workers with desk-jobs, they too have to work around deadlines, manage their backlogs and consequently work long hours in the office if needed. Those who work in fields that are related to their educational background like software testing, law book indexing, and manpower recruitment (as shown in Table 11.3) think the job helps to enrich their skills. Those with a background in a different field think that they learn something new on the job, with IT skills and process documentation most frequently mentioned.

Comparing the perception of currently employed workers in three sub-segments (Table 11.4) we find that although the majority of them find that work is demanding, slightly more call centre agents (11 per cent) than back-office support workers (4.5 per cent) perceive their job as totally undemanding intellectually. More generally, however, results of the survey indicate that most currently employed workers think of their job as moderately (37 per cent) to highly intellectually demanding (46 per cent). Hendriks (2010), found similar results among BPO workers in Baguio, Philippines and suggested that doing unfamiliar work leads to a perception that work is difficult. In the same vein, it is well possible that the challenging performance parameters call centre and back-office support agents have to negotiate, help to fuel workers' perceptions that their job requires much of one's intellectual capacities. There are call centres that raise the metrics as soon as workers reach them and in so doing the firms are able to maintain high levels of performance relative to competitors (female, HR manager, 6 Feb 2014). As a former call centre agent shared:

> You will not be able to meet the target of [ABC] because it always goes up. For example, if the BPO third-party hits the target of [ABC] today the following day it will be higher. It cannot be the same, it has to be higher.
>
> (Female, former call centre agent, 25 January 2014)

Workers, in result, are continuously on their toes. Thus while work can be perceived as monotonous and dull, there is constant pressure for efficiency and effectiveness.

Table 11.4 Perceived level of intellectual demand of current job in call centre, back office and software, n=358 (0-NR; 1-totally undemanding; 2-low; 3-neither; 4-moderate; 5-highly demanding)

BPO sub-segment	*0*	*1*	*2*	*3*	*4*	*5*	*Total (%)*
Call centre	1	17	7	6	52	71	154 (43)
Back office		9	9	11	82	89	200 (56)
IT/Software					1	3	4(1)
Total (%)	1(0)	26(7)	16(4)	17(5)	135(38)	163(46)	358(100)

Skills development on the job

Basic experiences in various BPO segments

Turning to skill development on the job, our survey results reveal that call centre agents and back-office support workers learn a range of similar skills suggesting that there is a large degree of commonality in terms of the skills used and demanded by these two BPO segments. Among the eight most frequently cited learned skills, seven are shared between call centre and back-office support workers (Table 11.5).

As noted, BPO firms consider only a small portion of applicants to be sufficiently skilled to be directly hireable. Upon hiring, a trajectory of potential skill development begins. Those newly hired tend to receive at least some additional training depending on the account. While specific practices differ, call centres usually start with two weeks of English Skills Training (ESL), two weeks of product training and a week of 'nesting' in which workers start to receive calls while

Table 11.5 Skills learned on the job by call centre and back-office support workers

Call centre workers (n=154)	*Frequency*	*Back-office support workers (n=200)*	*Frequency*
*Most frequently mentioned learned skills**			
Communication	133	Communication	161
Problem solving	101	Flexibility	128
Cooperation with others	98	Analytical thinking	116
Customer orientation	95	Problem solving	115
Flexibility	94	Cooperation with others	115
Self-control and stress-resistance	80	Cultural sensitivity	101
Self-confidence	76	Self-control and stress-resistance	101
Analytical thinking	73	Customer orientation	93
Least frequently mentioned learned skill		*Responses by call centre and back-office support workers combined*	
Knowledge in CAD/CAM, drafting, modelling, product design		7	
Appraisal and control of quality of raw materials, semi-products and products		10	
Administration of manufacturing and operational records		12	
Control of production processes and product parameters		23	
Basic competencies in science and technology		23	
Journalism and writing skills		25	
Environmental awareness		28	
Knowledge in legal applications		37	

Note: * Respondents were given a list of 40 different skills and asked to signal which skills (max. 10) they gained from working in the BPO sector.

still getting some coaching (male, call centre agent, 10 April 2013). One big captive centre says that they no longer give English language training because they only hire experienced workers (female, HR manager, 6 Feb 2014). Meanwhile, third-party BPO firms shoulder the costs of training for clients. A former vendor manager explains:

> So, for example, if I (client) tell them (third-party BPO firm) I need 20 additional agents to raise the headcount from 80 to 100, I will pay for the training of those 20 agents. But, if suddenly 20 persons out of the hundred resign the BPO firm will pay for the training of the new 20. That's why retention is a big strain on firms because they pay for those training courses.
>
> (Female, former vendor manager, 30 October 2013)

For back-office work, the intensity and duration of training depends on the account and needs of the company. In the survey, we asked workers whether they attended training in the company prior to the start of actual work and found that 75 per cent did (with training lasting a month). The fact that 25 per cent did not suggests that pre-work training is not a standard practice across the sector. Part of the reason may be because some firms are cutting-down on training expenses by only hiring experienced workers. The following excerpts from two informants in offshore recruitment, the first from a big multinational firm and the second from a small firm providing legal data services, provide a sharp contrast:

> I was not given formal training because when I came in they were badly in need of help so for my training I was simply asked to 'shadow' a colleague and then I picked up from there. My boss at that time taught me one-on-one, but there was no formal training.
>
> (Female, former BPO worker, 12 Aug 2013)

> We studied US laws for six months. We did nothing but training – US Supreme Court, state courts, they taught us those. After six months we trained how to analyse court decisions, how to link it with court decisions.
>
> (Female, former BPO worker, 4 Mar 2013)

Once workers start for real, existing skills are sharpened and new skills acquired via day-to-day learning by doing. Below, basing ourselves on both our survey and interview data, we examine how workers' skill portfolios are broadened and/or enhanced through working in the industry.

Skills development through job-related training

In the survey, we asked workers to cite at least two company-provided training courses that they had attended in the last 12 months. Almost half of them (47 per cent) reported that they had indeed attended training in said period. The results indicate that training topics fall under one of seven types: leadership;

communication; self-enhancement; IT; product training; customer orientation; and process improvement (Table 11.6).

Communication skills are most frequently cited as having benefitted from working in the BPO sector. Analysis of the interview narrations reveals that the manner of speaking and the delivery were believed to have improved most and workers become more skilled in getting their message across more effectively. Among others, workers reported that the job helped them to overcome stammer, moderated their strong regional accents, and helped them to acquire a more neutral accent. Respondents also developed higher self-confidence in speaking, which may be related to their perception of self-efficacy on a job where they need to interact with foreign customers. One respondent reflects on his experience as follows:

> [Call centre work] can somewhat help to build your confidence, even if you don't see the person you are talking to. You may also learn to be argumentative especially when you have to stand your ground with American or British callers. In the Philippines, our culture is if you can speak with [foreign, i.e. American rather than Filipino or regional] accent, even though your grammar is not perfect, people will think you're smart – which is not necessarily (true). So in a way the call centre helps because when you acquire an accent as some people do (laughs) you gain confidence, you will be more positively regarded by others and that can help you to pass future interviews (with non-BPO employers).
>
> (Male, former call centre agent, March 6, 2013)

Table 11.6 Training programmes followed by BPO workers divided by themes (n=167)

General topic	*Responses*	*Examples (responses)*
Leadership	35	Leadership training (12); project management (6); management training (3); front line leadership (1); online mentor management (3); supervisory leadership (4)
Communication	29	Business communications (2); business writing (3); presentation skills (2); communication skills (9); English proficiency (4); speak smart (4)
Self-improvement	26	Seven-habits of highly effective people (7); assertiveness training (2); Six Sigma (9); French classes (1); Spanish classes (1)
Information Technology	25	Excel (11), PowerPoint (1), Oracle (2), technical software (2); SAP (1); MIS (1)
Product training, refreshers, values	21	Product training (8); up-training (2); core values (3); Corporate etiquette (2)
Customer service orientation	18	Service culture (2); beyond service excellence (6); customer service (5); stakeholder engagements (2)
Process improvement	13	Information Technology Infrastructure Library (ITIL) (5); process improvement (5)

Survey results also show that workers perceive that they develop flexibility. Flexibility is an important survival skill for BPO workers: one must learn to adjust one's body (that is to a nocturnal work rhythm) and condition the mind to temporal parameters that circumscribe and define the job. Respondents view the BPO work environment as highly dynamic: work shifts may change every (other) week, accounts may close at short notice, firm management may change, and relationships with teammates can turn sour. In this respect it is not surprising that workers also think they develop self-control and stress-resistance.

Next there is the industry's almost obsessive preoccupation with time: tolerance for tardiness and absences is low, prompt return after every break is a must, and there is always the average handling time (AHT) that must be met. Workers say that not a few companies go to the extreme (for example by recording both frequency and time spent in the toilet), but they also admit that this kind of culture helps to develop a work ethic oriented towards efficiency and punctuality – valued qualities in practically any part of the labour market.

Workers also learn to cooperate with others. While the work outputs are by and large credited to individuals, workers are placed in different teams and have to cooperate with members to raise 'team stats'. If a member is lagging behind in his or her AHT, one or two other members who are more seasoned may 'buddy-up' with her or him and take some of the calls so as to log shorter AHTs for the member. The buddies may also listen to the calls and coach the member by showing areas that he or she can improve on, for instance the avoidance of filler words like 'actually' which can add to the handling time. Members also become conscious of their own and their team's attendance record as this could likewise drag team scores. Best performing teams for the month receive a number of incentives like getting choice shift schedules, monetary rewards, or special treats. One of the benefits of having people from different fields and backgrounds working together is that interaction fosters learning among peers. Six interview respondents remarked that they learned to solve problems and acquired other skills with the help of co-workers equipped with alternative knowledge and skills obtained from different educational or work experiences. In addition, having younger and older workers sharing a workspace was reported to help building mentor-mentee relationships with younger workers sometimes teaching the older ones, for instance in technology related matters, and the older ones teaching the younger about more practical life-related concerns.

In addition, workers develop problem-solving skills which show in their understanding of customers' issues and their mastery of the account. While most jobs are routine tasks, workers from time to time deal with unfamiliar questions. Extensive knowledge of the system and 'knowing where to look' are required to deal with these. Shifting to a new system or account expansion can also call on workers' problem-solving skills.

Another skill that workers reportedly develop is customer service orientation. For many BPO firms customer service is core business; workers are mustered to provide 'excellent customer service' which is measured through customer satisfaction ratings (CSAT). To ensure this, some foreign clients maintain their own company representative in offshore sites to monitor the performance of call centre vendors and their adherence with the terms of the Service Level Agreement (SLA).

Finally, interview results indicate that especially back-office workers also learn to use new business software, process documentation and data analysis. These skills are hardly mentioned by call centre agents, for whom communication skills are key. As two former back-office workers note:

> My skills in using (MS) Excel were enhanced on the job. In college, we studied (MS) Excel but I was able to really apply it and improve on it on the job, like using shortcuts, making tables, etc... The most helpful skill for me was knowledge on how to use SAP.
>
> (Female, former BPO worker, 14 March 2014)

> The skill to manipulate tools that we use, like our company's databases – because [my current job] has something to do with databases […]. Even if it is not part of the process, […], they encourage that. So the manipulation of databases and doing things by yourself – I got those and I think they were very useful. Also, looking for resources from nothing – your analytics within the account, the resourcefulness basically. I also picked up professional communication like I did not have any idea about how to compose an email in a business context and I picked that up as I went along.
>
> (Female, former BPO worker, 12 Aug 2013)

In back-office support performance is measured by means of output and while in some accounts output is easily determined by the number of purchase orders released, collections made, or data entered in a system, some workers must keep a log of outputs. According to two analysts who used to work in insurance and offshore recruitment firms:

> [I learned] records management, like I said in company A (data management) and company B (insurance company), we kept records, particularly of accounts that we have worked on. We kept records on how long we have worked on this, what they call productivity rate. And of course I also learned financial skills, because in the two companies I mentioned, we did many calculations.
>
> (Male, former back-office worker, 18 January 2013)

> The specific process skills that I learned on the job such as how to make a purchase order, I cannot recall that anymore but what did stick with me is the culture that you always have to plan, look for lots of opportunities all the time. In the conduct of my job in QA recruitment I learned the discipline to always check for accuracy and when you deliver it has to be on time, perfect every time, always efficient – very idealistic attitudes at work.
>
> (Male, former back-office worker, 12 January 2013)

While workers undoubtedly gain domain knowledge (usually related to the client account(s) they maintain), this knowledge does not necessarily facilitate a transfer to a non-BPO job or a business start-up. Little evidence was found that workers acquire enough domain knowledge to make them confident of landing employment in the industry of the client itself (for instance, someone managing certain tasks for an American law firm being able to land a job at an US-based law firm). Table 11.7 shows that with a view to the future, many workers actually prefer to get into a job that is related to their field of training or education, although there is also a good number seeing a prolonged stay in the BPO.

Interestingly, workers who trained in specific fields like education or nursing express concern that BPO work may cause them to lose previously acquired domain knowledge. One nurse-turned BPO worker resigned from a call centre job after six months because 'I felt that time is running out and I needed to start training as a nurse' while a teacher-turned-call centre agent worried whether she could still qualify as a teacher after having worked as an agent for almost a year (female, call centre agent, 16 Aug 2013). Two HR managers express different views on this issue. According to one, the longer a person works in BPO, the lower his chances are to return to a previous field, while the other thinks that a career shift should be possible too: 'If you're in financial services and you're doing accounts payable in different countries, accounts payable has a universal language, so if you have experience in a BPO company you can move to any company doing accounts payable' (female, VP HR, 6 Feb 2014).

Meanwhile, there is little evidence from the interviews that BPO work inspires or enables workers (current or former) to venture into entrepreneurship. As noted, workers do not appear to gain deep domain knowledge of the industries they engage with nor do they acquire a network that enables them to access opportunities to start a business. As one former BPO worker remarked:

> Everyday your mind-set is interacting with people, teammates, passion for excellence, customer service – it revolves around that so your creativity is restrained by the routine. If you tell your team leader about a business idea,

Table 11.7 BPO workers' next job preferences (n=358)

If you were to leave this company next month, what kind of job or activity would you apply or aim for given your educational background and set of occupational skills now?	*%*
A job related to my field of training or educational background	30
Still in IT-BPO	24
Start a business	16
An overseas job	15
A job similar to this one but not IT-BPO	10
A government job	3
Any job as long as not IT-BPO	2

> you will get indirect discouragement because you might resign from your job. S(he) will say, 'good, wonderful idea, but you should also weigh the consequences'. If you talk about it with your teammate, (s)he will say 'that's tough. I think working here and getting paid every month is cool. This is okay'.
>
> (Male, former back-office support worker, 13 May 2013)

On a different level some firms do allow employees to unfurl entrepreneurial activities in the workplace, usually in the form of petty trade – selling snacks, watches, clothes, and cosmetics to fellow workers sometimes on loan, a practice that creates a bond between lender and borrower and which helps promote retention.

Simultaneously, few workers appear willing to invest in employability-increasing personal development activities. Only 2 per cent of the workers said they invested in self-enriching activities such as enrolling in further studies, learning a foreign language, attending different seminars, taking review sessions or refresher seminars in their field of study.

Conclusions

This chapter has dealt with the question whether and how services offshoring contributes to human capital development in developing countries. It has done so by examining how the BPO industry in Metro Manila utilises the skills of its workers and how working in the industry enhances workers' skill portfolios.

Our findings suggest that workers generally hold positive views on how BPO work contributes to skill development and the building of work ethic. Work in the BPO industry seems to notably enhance and add to workers' generic skills. Communication skills benefit a lot and workers learn to cooperate, be flexible, efficient, punctual, accurate and self-reliant in problem solving. A lucky few gain management skills. In addition to these, back-office support workers may gain some analytical and software application related technical skills. As such they seem to acquire a slightly greater variety of skills than call centre workers. Most workers also develop some degree of domain knowledge, dependent on the type of account(s) they manage. Variation here, however, is great with some accounts requiring only limited or superficial understanding of the account matter (e.g. general customer care), while others require quite specific or deep knowledge of the matters at stake (e.g. legal services, user support). With exception perhaps of the latter, the skills so gained through BPO work seem well portable or transferable to other work environments. Most if not all skills acquired help to make workers more versatile and confident, and thus should contribute to workers' employability and potential productivity. Whether and to what degree the newly acquired skills are actually and productively used in new work contexts, of course depends on the particularities of the work environments ex-BPO workers are moving to (see also Marasigan, 2015).

Yet, a downside was identified as well. Working in the BPO appears to hardly appeal to workers' previously (mostly through study) acquired technical skills (e.g. medical, engineering, teaching skills). Especially when people stay in the BPO sector for longer periods of time, such technical skills may erode and domain knowledge may grow obsolete. A consequence of this would be that such workers, in spite of having enhanced and broadened their generic skills, end up being less qualified to work in their fields of previous study. With the industry (in Manila) indeed taking its workers from the most highly educated segments of the labour market, worries about the rapid growth of the industry creating a situation where much (in the local context) high quality human capital remains under-utilised and is made (at least temporarily) unavailable to other economic sectors, seem justified.

Our findings lend at least partial support to previous work by Beerepoot and Hendriks (2013) and Vira and James (2012) who also found that people working in the offshore services industry do accumulate new skills and manage to enhance their employability. Our observation on the other hand that the industry leaves many of workers' previously acquired technical skills underutilised and may foster misallocation of human capital in the economy, relates to the more sombre evaluations of the industry's contribution to employment and skill development by Bird and Ernst (2009) and Taylor and Bain (2005).

With the current research having focused on the BPO segment of the offshore services industry, there is scope for future research into the skills utilisation and skills development effects in the ITO and KPO segments. With the latter representing the more skills and knowledge intensive services offshoring segments, outcomes may be different from the ones presented here. In addition, to establish more clearly the degree to which skills acquired in the offshore services industry help to enhance the productive capabilities of a local workforce and the competitiveness of a local economy, more research is needed tracking the employment and career trajectories of former BPO (and ITO and KPO) workers (see also Marasigan 2015). Finally, it would be interesting to engage in international comparative research to see how outcomes are affected by for instance local educational and labour market conditions.

Note

1 Reliable, exact numbers are not available. Frequently cited is the 4.1 million offshore service jobs as estimated by McKinsey for 2008 (see e.g. Fernandez-Stark *et al.*, 2014). More recently, in 2012, India and the Philippines, the two largest recipients of offshored services jobs, accounted for almost three million jobs between them (Kleibert, 2015).

References

Barro, R.J. and Lee, J.-W. (2010). A New Data Set of Educational Attainment in the World, 1950–2010. NBERWorking Paper No. 15902. Available at: www.nber.org/papers/w15902 [accessed on 22 October 2015].

Batt, R., Holman, D. and Holtgrewe, U. (2009). The Globalization of Service Work: Comparative Institutional Perspectives on Call Centers: Introduction to a Special Issue of the "Industrial & Labor Relations Review". *Industrial & Labor Relations Review*, 62(4), pp. 453–88.

Beerepoot, N. and Hendriks, M. (2013). Employability of Offshore Service Sector Workers in the Philippines: opportunities for upward labour mobility or dead-end jobs? *Work Employment Society,* 27, pp. 823–41.

Beerepoot, N., Lambregts, B. and Kloosterman, R.C. (2015). Conclusions: offshore services and the road to development. In: B. Lambregts, N. Beerepoot and R. C. Kloosterman (eds) *The Local Impact of Globalization in South and Southeast Asia: Offshore business processes in services industries.* London and New York: Routledge, pp. 208–14.

Bird, M. and Ernst, C. (2009). Offshoring and Employment in the Developing World: Business process outsourcing in the Philippines. International Labour Organization: Geneva.

Bottini, N., Ernst, C. and Luebker, M. (2007). Offshoring and the labour market: what are the issues? International Labour Organization: Geneva.

Chadee, D., Raman, R. and Michailova, S. (2011). Sources of Competitiveness of Offshore IT Service Providers in India: Towards a Conceptual Framework. *Competition & Change*, 15(3): pp. 196–220.

CHED (2004). CHED Memorandum Order (CMO) No. 30 Series of 2004. Retrieved from www.ched.gov.ph//wp-content/uploads/2013/07/CMO-No.30-s2004.pdf [accessed 4 April 2015].

CHED (2012). CHED Memorandum Order (CMO) 34 Series of 2012. Retrieved from www.ched.gov.ph/wp-content/uploads/2013/07/CMO-No.34-s2012.pdf [accessed 4 April 2015].

Di Gropello, E. (2010). Skills for the Labor Market in the Philippines. The World Bank.

DOLE (2012). Industry Career Guide: Business Process Outsourcing. Philippines: Department of Labor and Employment.

ECLAC (2008). Foreign Direct Investment in Latin America and the Caribbean. Santiago: Economic Commission for Latin America and the Caribbean.

Fernandez-Stark, K., Bamber, P. and Gereffi, G. (2010). Workforce Development in Chile's Offshore Services Industry. Durham: Duke University, Center on Globalization Governance and Competitiveness. Commissioned by CORFO.

Fernandez-Stark, K., Bamber, P. and Gereffi, G. (2011). The Offshore Services Global Value Chain: Economic Upgrading and Workforce Development. Center on Globalization, Governance and Competitiveness (CGGC), Duke University.

Fernandez-Stark, K., Hernández, R.A., Mulder, N. and Sauvé, P. (2014). Introduction. In: Hernández, R.A., Mulder, N., Fernandez-Stark, K., Sauvé, P., López Giral, D. and Muñoz Navia, F. (eds). *Latin America's Emergence in Global Services. A new driver of structural change in the region?* Santiago, Chile: United Nations publication, pp. 13–37.

Gereffi, G. and Fernandez-Stark, K. (2010). The offshore services value chain. Working Paper 5262. The World Bank.

Graf, M. and Mudambi, S.M. (2005). The outsourcing of IT-enabled business processes: A conceptual model of the location decision. *Journal of International Management*, 11(2), pp. 253–68.

Hayter, S. (2004). The social dimension of global production systems: A review of the issues. International Labour Office: Geneva.

Hendriks, M. (2010). Local Outcomes of the Changing International Division of Labour: The Emergence of the Offshore Service Sector in Baguio City, the Philippines. University of Amsterdam.

Hernandez, B. (2012). Addressing the Labor-Market Mismatch in the IT-BPO Industry (Presentation, AIM, Makati City, October 5, 2012). Presented at the Addressing the Labor-Market Mismatch for Global Competitiveness, AIM, Makati City.

Hollinshead, G., Capik, P. and Micek, G. (2011). Beyond Offshore Outsourcing of Business Services. *Competition & Change*, 15(3), pp. 171–76.

Inkson, K. and Arthur, M.B. (2001). How to be a successful career capitalist. *Organizational Dynamics*, 30(1), pp. 48–61.

International Labour Organization (ILO) (2014). World of Work Report 2014: Developing with Jobs. International Labour Organization: Geneva.

James, A. and Vira, B. (2010). "Unionising" the new spaces of the new economy? Alternative labour organising in India's IT Enabled Services–Business Process Outsourcing industry. *Geoforum*, 41(3), pp. 364–76.

Kleibert, J.M. (2015). Services-led economic development. Comparing the emergence of the offshore service sector in India and the Philippines. In: Lambregts, B., Beerepoot, N. and Kloosterman, R.C. (eds). *The Local Impact of Globalization in South and Southeast Asia Offshore business processes in services industries*. London and New York: Routledge, pp. 29–45.

Magtibay-Ramos, N., Estrada, G. and Felipe, J. (2008). An input–output analysis of the Philippine BPO industry. *Asian-Pacific Economic Literature*, 22(1), pp. 41–56.

Marasigan, M.L. (2015). How work in the BPO sector affects employability: perceptions of ex-BPO workers in Metro Manila In: Lambregts, B., Beerepoot, N. and Kloosterman, R.C. (eds). *The Local Impact of Globalization in South and Southeast Asia Offshore business processes in services industries*. London and New York: Routledge, pp. 138–52.

Oshri, I., Kotlarsky, J. and Willcocks, L.P. (2011).*The Handbook of Global Outsourcing and Offshoring*, 2nd edition. London: Palgrave Macmillan.

Sullivan, S.E. (1999). The changing nature of careers: A review and research agenda. *Journal of Management*, 25(3), pp. 457–84.

Taylor, P. and Bain, P. (2005). 'India calling to the far away towns': the call centre labour process and globalization. *Work, Employment and Society*, 19(2), pp. 261–82.

Tholons (2014). 2014 Tholons Top 100 Outsourcing Destinations: Regional Overview. [Online] Available at: www.tholons.com/TholonsTop100/pdf/Tholons_Top_100_2014_Rankings_and_Report_Overview.pdf [accessed 7 April 2015].

UNCTAD (2004). World Investment Report 2004: The Shift Towards Services. UNCTAD: Geneva.

Vira, B. and James, A. (2012). Building Cross-Sector Careers in India's New Service Economy? Tracking former call centre agents in the national capital region. *Development and Change*, 43, pp. 449–79.

Visconti, K. (2012). Beyond English: the key skills BPOs need [Online] Rappler. Available at: www.rappler.com/business/13699-what-key-skills-are-needed-to-work-in-bpo [accessed on 6 March 2015].

Wadhwa, V., De Vitton, U.K. and Gereffi,G. (2008). How the Disciple Became the Guru: Is it Time for the U.S. to Learn Workforce Development from Former Discipline India? [Online] Durham, NC: Duke University and the Kauffman Foundation. Available at: www.soc.duke.edu/GlobalEngineering/papers_disciple.php [accessed on 27 October 2015].

12 Beyond call centres

The indirect effects of service sector-driven growth on new middle-class consumption and retail in Mumbai

Panos Papadongonas and Niels Beerepoot

Introduction

The growth of the Indian ICT-ITES sector has been the subject of extensive research in the past decade (Agrawal *et al.*, 2012; Bhattacharjee, 2012; Agrawal, 2014; Bhattacharjee and Chakrabarti, 2015; Kleibert, 2015; Beerepoot and Kumar, 2015). Research shows that India is the first large but extremely poor country that has entered the global economy on the basis of the labour power of its middle-class white-collar workers (Dossani and Kenney, 2009, 99). The Indian ICT-ITES sector has been instrumental in providing direct employment to around three million workers (NASSCOM, 2010). In addition, NASSCOM (National Association of Software and Services Companies) claims that every job in the ICT-ITES sector indirectly generates four more jobs (ibid.). As such, the ICT-ITES sector has contributed to transforming the economy, driving a shift in spending power and stimulating the rise of a new middle class (NMC) (Fernandes, 2000; Fernandes and Heller, 2006; Nijman, 2006; Das, 2009; Murphy, 2011).

Various consultancy and policy reports emphasise how the NMC above all should be seen as a consumer force (McKinsey, 2007; Goldman Sachs, 2008; ADB, 2010; Credit Suisse, 2010, 2011). Advertisers and investors allegedly have great hopes for the role that the NMC can play in boosting global consumption (Trentmann, 2010). Service-outsourcing professionals form a significant component of this NMC in emerging economies (see Fernandes, 2000; Upadhya, 2008). Undoubtedly, the ICT-ITES sector stimulates new consumer expenditures among its overwhelmingly young professionals by providing higher disposable incomes. As a consequence of their simultaneous exposure to foreign media and the incurred cultural influences through their internationally-oriented work, service-outsourcing professionals have become more aware of the lifestyle and consumption trends worldwide. Consumption as a core component of lifestyle is ultimately related to complex processes of self-identification (Giddens, 1991; Warde, 1994; Friedman, 1995; Nadeem, 2009). Furthermore, NMC consumption, especially in its conspicuous form, influences the evolution of local retail landscapes, the implications of which extend to local economies and how these economies act as nodes in the global economic nexus.

Using empirical data collected from service-outsourcing professionals in India's economic centre, Mumbai, in early 2014, this chapter examines how service-driven economic development goes beyond economic indices alone. The academic contribution of this chapter is threefold. First, it contributes to the existing literature on NMC formation by offering an empirical account of the process of the formation thanks to service outsourcing in Mumbai. Second, by identifying how identity is conceptualised in different segments of the service-outsourcing sector, it builds on Kirk and Wall's (2011) theoretical framework concerning the relationship between work and identity. The value of this approach becomes even greater if we consider that there is already significant evidence about the lifestyles of the professionals who participate in the call centre or low-end business process outsourcing (BPO) branch (Hubacek *et al.*, 2007; Pal and Buzzanell, 2008; Upadhya, 2008; Nadeem, 2009; Murphy, 2011; Yong Jin, 2011), but only limited research on professionals in more complex service-outsourcing operations, such as knowledge process outsourcing (KPO). Third, by 'mapping' NMC consumption and the rationale behind it, this chapter offers insights for retail marketing strategies in an emerging market, represented by Mumbai.

The next section provides the theoretical framework for this chapter. It commences with a brief description of service outsourcing and how it has contributed to the emergence of a NMC in the Global South. The third section provides a description of the research methodology. The fourth section concentrates on the testimonies of service-outsourcing professionals: how do they actually consume and how does that consumption relate to the complex interaction between their service-outsourcing careers and processes of identity formation? The fifth section looks at service-outsourcing-derived consumption from a retailers' perspective and especially the local (luxury) market's strategies to serve the new needs of the NMC. The sixth section summarises the main points of this research and provides an agenda for further research.

Theoretical framework

Service offshoring and the emergence of the new middle class

Offshore outsourcing of services entails transferring tasks to another supplier in the value chain, notably located in the Global South, in order to take advantage of lower costs (Gereffi *et al.*, 2005; Lambregts *et al.* 2015). While service outsourcing began as the transfer of secondary and low-level operations to lower cost countries, today high-level operations are also outsourced, following the increased complexity of economic transactions worldwide (Agrawal *et al.*, 2005). The former is generally termed BPO and the latter is labelled KPO, but Information Technology (IT) operations also occupy the high-end of the spectrum. BPO can be broadly decomposed into three areas of functional operations: 1) business administration, 2) supply chain management and 3) sales, marketing and customer care (PwC, 2005). KPO, which took off after 2003, requires significant domain expertise (for example, market and business research) (Evalueserve, 2007). KPO firms function at a higher level and they can be differentiated from BPO firms in terms of greater requirements of domain expertise, size and client involvement in the execution process (Evalueserve, 2007). Important activities within the KPO branch include: 1) banking, securities and industry research services, 2) contract research organisations and bio-pharmaceutical services, 3) data management, mining, searching and analytics and 4) legal, paralegal and intellectual property support services (ibid.).

In India, the ICT-ITES sector offers employment to young graduates, irrespective of their social background, so that they do not have to rely on emigration or preferential treatment practices to secure their future through government jobs (Rastogi and Pradhan, 2011; Krishnan, 2015). Allegedly, 40 per cent of the people joining the outsourcing industry in India 'are first generation graduates, who come from socially disadvantaged backgrounds' (Rastogi and Pradhan, 2011: 25). The general increase in job opportunities and incomes and the increased exposure to the rest of the world by means of careers in the ICT-ITES sector have undoubtedly influenced social processes in India, leading to the creation of a NMC (Krishnan, 2015). Fernandes and Heller (2006: 497) argue that the NMC 'is a tangible and significant phenomenon, but one whose boundaries are constantly being defined and tested.' Das (2009) describes India's NMC as ambitious, globally-informed and dynamic. Unlike the old middle class, which was based on British education practices, the NMC identifies itself on the basis of monetary resources, consumption and efficiency (Krishnan, 2015). The NMC is evidently developing its own identity which is a departure from previous class structures in India.

New middle-class identity, consumption and the retail sector

Identity formation processes take place at both the individual and the collective level (Yoon, 2011). The social group to which an individual belongs shapes the person's self-image and, via interacting with his or her kin, self-esteem is reinforced and 'others' are defined. Therefore, self-identification is a relational process which may be described as 'an ongoing becoming that arises as an emergent product of comparison' (Koefoed and Simonsen, 2012: 625). Consumption patterns also constitute a form of identification; consumers purchase conspicuous goods to satisfy not only material needs but also social and psychological needs in order to attain prestige (Amaldoss and Jain, 2005; Hudders, 2012). Similarly, Holt (1995) and Vigneron and Johnson (2004) claim that the motivation behind consuming luxury goods lies primarily in their symbolic value. Engaging in luxury consumption symbolises pursuing exclusivity from the masses and at the same time conforming to the expectations of the kin (Amaldoss and Jain, 2005). Among the better-educated and internationally-exposed members of the NMC, luxury consumption can then function as the means to differentiate themselves from mainstream society. Evidence suggests that Veblen effects – the fact that decreasing the price of a product minimises people's interest in buying it – may be empirically significant in the luxury goods market (Bagwell and Bernheim, 1996). Veblen effects arise from a consumer desire to signal uniqueness, wealth and status, in which case luxury consumption acquires a conspicuous character. Once the uniqueness aspect of owning a certain product diminishes, a snob effect (ignoring the product) may take place (Corneo and Jeanne, 1997).

In the existing literature, most of the socioeconomic implications of the service-outsourcing sector, such as the intensification of extravagant consumption practices, are mainly associated with call centre or BPO jobs (Krishnan, 2015). Nonetheless, Upadhya (2008) considers that differences in consumption patterns and preferences are discursively employed by more skilled professionals to distinguish themselves from the young conventional call centre crowd, whose frivolous lifestyles and dissolute consumption habits they disdain. Similarly, encouraged by their higher incomes, the highly skilled and better-paid BPO, IT and KPO professionals may consciously or unconsciously attempt to differentiate themselves from the lower-paid call centre BPO professionals, which may translate into differing consumption patterns.

Corneo and Jeanne (1997: 68) suggest that 'an individual's status depends on public perceptions about his relative income, and the individual chooses his consumption pattern by trading off the gain in status obtained by impressing the public with the loss in the consumption of commodities that are intrinsically more useful.' This points to a move away from utilitarian motives towards identification processes through (conspicuous) consumption, which in turn points to a positive attitude towards cultural change (Dubois and Duquesne, 1993). With regard to cultural factors, Dubois *et al.* (2005) and Warhurst and Nickson (2009 in Murphy, 2011) suggest that luxury items are present in all societies, but the role they play may be culturally determined. As Clammer (2003: 403 in Murphy, 2011) notes, 'consumption behaviour [...] both creates the primary cultural context in which the new middle classes in Asia operate and provides the linkage between globalization and local urban cultures.' Similarly, Murphy (2011) argues that luxury consumption is an expression of how the NMC has adopted transnational values and lifestyles.

The conspicuous consumption of the NMC adds a new, international dimension to the retail environment of emerging economies (McKinsey, 2013a; 2013b). Consumption is often viewed as a manifestation of social and financial well-being (McKinsey, 2013c). As the NMC in emerging markets is growing exponentially, so are opportunities for the retail sector in these countries (Ernst and Young, 2013). The perception that the NMC needs in emerging markets are simple and that people's relatively lower income limits their purchasing aspirations is flawed (D'Andrea *et al.*, 2004). Furthermore, taking into account local socioeconomic conditions and consumers' aspirations is necessary for retailers who wish to successfully enter emerging markets (PwC, 2015). The NMC is evolving rapidly and the local retail markets will have to do so as well if they are to reap the benefits of this evolution.

Research methodology

This study is based on semi-structured interviews with BPO, IT and KPO professionals. The differentiation between these groups follows Kirk and Wall's (2011) process of socialisation into different work cultures and the way this process shapes the relationship between work and identity. The interviews included three components: the respondent's personal background (upbringing, family and work situation), (luxury) consumption aspects (description and motivation) and identity formation (how consumption becomes a form of identification). Furthermore, interviews with retailers took place in order to attest the retail sector's hitherto little researched perspective on the emergence of the NMC, partly as an effect of service-outsourcing operations in Mumbai, and the way this emergence led to consumption shifts in Mumbai. These interviews focused on store/mall information, current consumption trends, the evolution of consumption in recent years and future perspectives. The interviews were transcribed, coded (open and axial coding) and analysed (direct content analysis involving data branching and data reassembling) with specialised software. Within this context, snowball sampling presented the most opportunities for successfully reaching the target population. The fieldwork took place between January and April 2014. This resulted in 23 interviews with professionals (8 BPO, 7 KPO, 7 IT and 1 NASSCOM representative; taking 35 minutes per interview on average) and 12 interviews with representatives of the local luxury market (mall managers and tourism sector representatives) that took 20 minutes on average.

Luxury consumption of service-outsourcing professionals

The 23 interviewed service-outsourcing professionals belonged to different age groups with their ages ranging from their early twenties to their mid-fifties. The respondents were overwhelmingly men, with only one woman participating. Regarding their education, all respondents had received at least college education, about one third of them possessed MBA degrees and two people had received a PhD title. Their positions in the service-outsourcing functional spectrum varied greatly as well: from call centre representatives and computer programmers to (global) HR heads and managing directors. Most of the interviewed service-outsourcing professionals expressed satisfaction with their current lifestyle. Some considered it the natural order of things to be able to afford luxury items at this point in their life, whereas others expressed their gratitude for the outsourcing sector as a gateway to a more comfortable lifestyle. However, some respondents, mainly employed in low-end BPO companies, stated that their salaries did not allow them to engage in luxury consumption to the extent they would have desired. These respondents were very young professionals who recently entered the sector.

During the interviews, the respondents were asked to specify the kinds of luxury products that they usually purchase. For the purpose of this research, 'luxury consumption' was defined as products and services that are not of primary importance for one's survival and the true value of which resides not in the functionality or quality, but in the price (Veblen, 1899). The answers of the respondents were coded and luxury items were extracted from their testimonies and inventoried according to how many times they were mentioned. Figure 12.1 offers an overview of the luxury consumption of service-outsourcing professionals. The size of each word indicates the frequency with which it was mentioned by the respondents.[1]

Branded clothing and shoes were by far the most popular category among the respondents. Younger respondents and many professionals in the IT sector demonstrated a tendency towards purchasing more casual, yet branded, clothing. Older respondents and professionals of the KPO branch – who were often higher up in the corporate hierarchy – were keener on formal clothing, such as suits. Gadgets and electronics was the second most cited product category. BPO and KPO respondents mostly focused on mobile phones, tablets and TV's, whereas IT respondents appeared more tech-savvy, spending significant amounts of money on more sophisticated items such as laptops with certain characteristics and

Figure 12.1 Visualisation of service-outsourcing professionals' consumption

computer or sound system enhancements. Apparel and accessories followed as third most popular category which included watches, jewellery, bags and perfumes. KPO respondents were most conscious about the consumption of such products, recognising, to a greater degree, their added symbolic status-enhancing value. On the other hand, IT respondents were much less inclined to mention such products, which is not surprising, given the often more casual attire of the IT respondents. Finally, recreation (luxurious holidays and exclusive restaurants) was also mentioned by a small group of high-ranking BPO and KPO respondents. These respondents made reference to spending substantial amounts of their income on vacations. They also had an academic background and had experienced a more upper-middle-class type upbringing.

Purchasing luxury products was often associated with making a style statement. The degree to which looking good translated into being stylish or trendy varied and it appeared to be a function of personal background, character traits (detected through participant observation and interaction with the respondents) and professional function. People who grew up in less affluent environments were generally less prone to equate 'nice' with 'trendy' or 'stylish'. Rather, they demonstrated a tendency towards stressing the 'feel-good' factor of branded products. The visual, hedonic aspect of luxury consumption among the interviewees appeared to be more related to age, character traits and specific function than their association with the BPO, IT or KPO branch per se. The following respondent testimonies illustrate the various motivating factors behind purchasing luxury goods:

> 'Looking good, feels good, so let's buy it!'
> 'It just has to look good for me!'
> 'The definition of 'nice' is [that] it has to be appealing to me.'
> 'I'll buy something which gives me a lot of pleasure, satisfaction, has a style statement, but not like [style] is my priority.'
> 'I work so much, I can afford it myself.'
> 'I am able to buy this because I have done something.'
> 'It boosts my confidence.'
> 'It makes you feel more secure. You don't need to watch your back and be seen poorer than you are.'
> 'Not to be left out.'

All respondents described a connection between the product and themselves. By wearing clothes that matched certain situations, such as working or celebrating, they were better able to perform a role that matched the situation at hand. Consciously, the service-outsourcing professionals make rational choices based on visuals and emotion. Subconsciously, however, luxury consumption acquires a more pervasive character, pertaining to the respondents' whole being. Clothing and other apparel do not only physically cover the body, but they act as the visible expression of the persona that the consumer wants to create at a given moment. This process of self-expression demonstrates a continuous interaction with the social environment and is at times strongly influenced by the social expectations that the respondents experience. The following quotes are characteristic:

> 'A connection, between the product and me.'
> '[Suitable clothes] give you confidence and they are related to how the body feels.'
> 'Your identity is going to come from the gadgetry, from the way you are.'
> 'As I grew into a manager position I realised you have to dress, because you make an impression.'
> 'It is kind of human nature.' [Talking about imitating his peers' purchases].
> 'It's like, if you are in the same flock, you do what everybody does.'

Overall, Holt's (1995), Nueno's (1998), Dubois', Czellar and Laurent's (2001) and especially Vigneron and Johnson's (2004) perceptions on the dimensions of luxury consumption were undoubtedly present in this research, as respondents clearly based their product selection on the same dimensions. The only possible exception related to the price factor of luxury items. Even though most respondents recognised that higher quality comes at a high(er) price, the majority of them had set certain limits on how much money they were willing to spend, therefore slightly negating the price dimension of luxury consumption, as suggested by Forbes (2011).

The retail sector response to new middle-class aspirations

The first mall in Mumbai opened in 1999 and since then there has been an explosive growth in the number of malls. Malls initially started campaigns to shape new entertainment practices and to make visiting malls a standard practice. They hired celebrities, such as Bollywood actors, for specially organised events in order to attract visitors. The experiment was deemed a success and the young, ambitious and fashion-seeking NMC started flooding the stores. A partial conclusion stemming from the retailers' testimonies is that visiting malls has become a noteworthy component of the daily life of a great part of the NMC. A retailer's testimony is characteristic: 'people are so used to malls, that if they don't go to a mall on the weekend, they feel they have missed out on something'. Responding to the question as to whether there have been shifts in consumption in the last few years, the vast majority of the retailers stated that clothing and apparel have remained the dominant product category in the luxury market. A more recent phenomenon is the impressive growth of the smartphone and gadget market, following global trends. Notably, the tourism industry is also booming because of the NMC. Retail and tourism representatives talked about a very promising market segment, which is globally-informed and rather demanding in their choices.

Perhaps the most important aspect of the consumption boom that is taking place in Mumbai is that the NMC is becoming increasingly aware of new brands, which is the principal finding of malls' own market research. Exposure to foreign cultures since India embraced globalisation has been the single most important factor of change, according to the retailers. The going-out culture was already part of Indian lifestyle and the proactive mall strategies to attract more clientele have been based on this and have added a new dimension to entertainment. According to one retailer, members of this NMC 'talk about: "I have been to this mall", "I am going to that mall tomorrow", they don't talk about "I've been to this zoo today, I am going to that park tomorrow".' This implies a major cultural shift towards a more consumerist approach to entertainment and personal fulfilment. The greater preference for malls is possibly a direct outcome of people's increased awareness about brands, but it may also be considered a pay-off of the malls' aggressive communication strategies during the last decade.

In the process of cultural transformation, the outsourcing industry has certainly been an influential factor. The interviewed retailers pointed to two aspects of the industry which they deem responsible for the observed changes in the market in the last few years. The first aspect relates to the nature of outsourcing activities, which bring professionals in daily contact with people from abroad. In numerous cases, professionals even get the chance to travel abroad and experience foreign markets. Upon return, word of mouth follows and the popularity of brands rapidly increases. Many interviewees stated that customers are already aware of products and brands that are not yet available in India. The second aspect relates to the significantly higher salaries in the outsourcing industry. The majority of the interviewed retailers strongly believe that the higher salaries have resulted in higher aspirations among the outsourcing professionals, rendering them a very promising segment of the NMC. Higher aspirations ultimately result in boosted self-esteem and self-identification through brands, a process which was evident in the professionals' testimonies and is confirmed by the retailer side.

> See, what has happened because of the outsourcing industry is that...a lot of people from the rural areas have come to the urban areas and the standard of living has increased substantially, [...] the expenditure part has increased especially and that has helped the organized retail market to grow. People have become conscious about brands [...] the outsourcing industry, they speak with people from the US or Europe. That is also one of the channels [of becoming brand-conscious].

Following the increased awareness of the Indian NMC clientele, the interviewed retailers seek to keep pace with these trends by employing a variety of strategies: continuous research, landscaping or zoning, and ensuring product availability and exclusiveness. Continuous research involves two approaches: obtaining the latest information about the newest trends internationally and conducting market research on the local customer preferences. Then the mall marketing teams attempt to find common threads and create a good match with the customer's expectations. At times, malls are prepared to experiment with products that have gained popularity abroad to test those propositions domestically. In a characteristic example, a mall manager stated that they periodically experiment with niche products, whether they be specialised delicacies or an exclusive houseware brand.

The second strategy that retailers often employ involves optimising zoning or landscaping. Zoning was described as the careful mixing of various stores so that they create an optimised visit to the mall by rendering product searching efficient. Landscaping relates to the combination of different mall functions, principally shopping and entertainment. When applicable, malls attempt to create a more relaxing environment by adding green and open spaces, so that customers can relax before continuing their shopping experience. Profitable premises, such as restaurants and amusement parks, are present in such areas.

The third aspect, product availability and exclusiveness, is highly relevant with regard to the NMC's aspirations. On the one hand, malls try to respond to their customers' demands to introduce new brands, with which their customers have become acquainted thanks to their international exposure. On the other hand, certain malls try to safeguard exclusiveness, by being the only ones to provide carefully selected brands or products. Such malls cater primarily to the upper end of the NMC spectrum. There are also other malls which try to create a more diverse profile in order to attract a wider social spectrum of consumers. A factor determining which of these two approaches the mall chooses to adopt relates to the actual location of the mall and its catchment area. Table 12.1 lists the malls and gives an overview of the number of outsourcing companies that fall within their respective catchment area (calculated with Geographic Information System software).

In the table, 'high' importance translates to service-outsourcing professionals being the majority of visitors, 'medium' translates to 25–50 percent and 'low' translates to less than 25 per cent, or a clearly expressed low importance in qualitative terms. In all malls the respondents recognised the contribution of

Table 12.1 Mall catchment area and perceived importance of service outsourcing professionals as visitors

Mall	*BPO/KPO/IT companies within catchment area*	*BPO/KPO/IT professionals as % of monthly visitors*	*Importance assessment*
RCity	51	20–25	low
Phoenix Market City	44	30	medium
InOrbit	35	unspecified	high
Infiniti	35	20	low
Hypercity	35	40	medium
Highstreet Phoenix & Palladium complex	29	unspecified	low
Viviana Mall	16	unspecified	low
Sobo Central	16	15–20	low

the outsourcing clientele. Such contribution varied from relatively low to very high.

When comparing the table data and the mall managers' testimonies, certain inconsistencies emerged. Notably, all visited malls were in the vicinity of outsourcing companies, though company density varied per catchment area. Nonetheless, there were managers who insisted that their malls do not serve the typical outsourcing crowd, but rather corporate crowds. This was especially relevant in South Mumbai, where outsourcing companies focus primarily on KPO operations. In the mind of certain mall managers, however, the outsourcing industry was clearly synonymous with BPO operations. This finding points to a possible normative incongruence, as the term 'BPO', representing the oldest service-outsourcing variant, has become synonymous with the outsourcing industry, whereas the more specialised and better-paid KPO operations may remain generally nebulous for the unfamiliar public. The general conclusion is that location is of high importance and it determines the entire strategy and profile of a mall. Based on this, malls in the sample can be classified into those targeting residential areas and those (partly) relying on the nearby outsourcing companies.

Conclusion

This chapter aimed to give an insight into the consumption practices of service outsourcing workers in Mumbai in order to understand the indirect effects of the rise of the ICT-ITES sector in this city. Not only does the expansion of the ICT-ITES sector stimulate growth of the retail sector, it also leads to new perceptions among its workers on luxury consumption and social status. Following Yoon's (2011), Koefoed and Simonsen's (2012) and Han *et al.*'s (2010) observations, the relational character of identity formation and the processes of comparison that are part of gaining or maintaining social status were prominent among the interviewed workers. Many respondents, especially those working in BPO and IT, traded utilitarian value for the cost of enhanced status when selecting products. This points to observations by Amaldoss and Jain (2005) and Hudders (2012) that purchasing conspicuous goods satisfies not only material needs, but also social and psychological needs, adding what Choo (2012) termed 'social value'. At the same time, the sample should not be treated as an integral group, since processes of differentiation from the materialist crowds were spotted among certain highly-educated KPO respondents. This corresponds to Upadhya's (2008) views, whereby older and more skilled professionals attempt to differentiate themselves from younger avid consumers, most notably call centre crowds.

However, it was not age per se, but rather the life phase of individuals which dictated how extensive their luxury consumption was. The divergent consumer behaviour of married respondents, who were saving for their children, suggests that Dubois *et al.*'s (2005) and Warhurst and Nickson's (2009 in Murphy, 2011) proposition that the role that luxury items play is culturally determined may hold true in the sample. Nonetheless, the general consumption findings point to a profound desire of all respondents to discover previously unknown luxuries, whether it be in the form of a smartphone, branded clothing or an exotic vacation. Therefore, Brosius's (2010) claims about a shift towards a more consumerist direction among the Indian NMC finds its application in the sample. However, taking into account the variety of the sample with regard to people's background and their motivation behind luxury consumption, it can be claimed that luxury consumption is a very important, but not unique, dimension of identifying oneself as a member of the Indian NMC.

Implications of ICT-ITES sector growth for the local retail sector

The views of the interviewed retail representatives on luxury consumption demonstrate certain similarities with the service-outsourcing professionals' testimonies in terms of the nature and motivation behind this consumption. The interviewed retailers appeared highly aware of the symbolic character of luxury consumption towards fulfilling certain emotional needs and are in accordance with existing theoretical perspectives on the motivation behind consumer behaviour (such as Dubois *et al.*, 2001; Vigneron and Johnson, 2004; Amaldoss and Jain, 2005; Hudders, 2012). The retailer side recognises the importance of the NMC clientele, with service outsourcing being considered an influential subgroup of this clientele. Consequently, the malls of Mumbai demonstrate a proactive behaviour to keep pace with the latest international trends and thereby satisfy the NMC's desire for previously inaccessible products and services.

In continuation of Nijman's (2006) claims – namely that the economic liberalisation of India has led to an explosion of demand for consumer durable goods – retailers expect the trend to continue. The ICT-ITES sector certainly has a role to play in this, as incomes continue to grow, allowing more people to join the NMC. As a result, the share of organised retail in Mumbai is set to increase from the current 7 per cent, even though mom-and-pop stores will continue to dominate volume-wise. Nonetheless, such an increase of organised retail will be slow. This offers a wider time frame to map and comprehend the potential clientele's needs and to devise strategies that will be more viable in the long run. Malls appear conscious of the importance of better comprehending the NMC's needs, as they continuously and carefully monitor their clientele and respond accordingly.

A possible misconception also became evident during this research. Among the interviewed retailers, the ICT-ITES sector was synonymous with the well-established BPO operations. However, IT and especially KPO operations were not viewed as service outsourcing. As a result, representatives of malls that were seemingly located close to IT and KPO organisations stated that their overall business model was not influenced by the service-outsourcing clientele, adopting a business model that treats customers solely in terms of income. The evidence stemming from this research about the role of the ICT-ITES sector in the consumption and retail landscape in India's economic centre highlights the necessity of additional research involving a greater and more varied sample. Such research should aim at further comprehending how the globalisation of economic activities – notably expressed by the ICT-ITES sector – interacts with local economies in the Global South and leads to processes of cultural hybridisation and socioeconomic transformation.

Note

1 The figure is a so-called tag crowd and was created using relevant online services (tagcrowd.com).

References

ADB (2010) *The Rise of Asia's Middle Class, Special Chapter, Key Indicators for the Asia and the Pacific*. Asia and Pacific. Manila: Asian Development Bank.

Agrawal, S., Khaitan, S., Shrivastava, S. and Banks, M. (2005) Destination India: Offshore Outsourcing and its Implications. *CTLR-Oxford*, 11(8), pp. 246–59.

Agrawal, S., Goswani, K. and Chatterjee, B. (2012) Factors Influencing Performance of ITES Firms in India. *Information Resources Management Journal*, 25(4), pp. 46–64.

Agrawal, S. (2014) Issues of Concern and Mitigating Strategies: A Study on ITES Firms. *Vision*, 18(3), pp. 205–16.

Amaldoss, W. and Jain, S. (2005) Conspicuous Consumption and Sophisticated Thinking. *Management Science*, 51(10), pp. 1449–66.

Bagwell, L.S. and Bernheim, B.D. (1996) Veblen Effects in a Theory of Conspicuous Consumption. *The American Economic Review*, 86(3), pp. 349–73.

Beerepoot, N. and Kumar, R. (2015) Upgrading Service Delivery and Employment Conditions Through Indirect Insertion in Global Value Chains. In: *Competition and Change*, 19(5), pp. 374–89.

Bhattacharjee, S. (2012) Efficiency Dynamics and Sustainability of the Indian IT-ITES Industry: An Empirical Investigation Using DEA, *IIMB Management Review*, 24(4), pp. 203–14.

Bhattacharjee, S. and Chakrabarti, D. (2015) Investigating India's Competitive Edge in the IT-ITeS Sector. *IIMB Management Review*, 27, pp. 19–34.

Brosius, C. (2010) *India's Middle Class: New Forms of Urban Leisure, Consumption and Prosperity*. New Delhi: Routledge.

Choo, H.J. (2012) Luxury Customer Value. *Journal of Fashion Marketing and Management*, 16(1), pp. 81–101.

Corneo, G. and Jeanne, O. (1997) Conspicuous Consumption, Snobbism and Conformism. *Journal of Public Economics*, 66(1), pp. 55–71.

Credit Suisse (2010) Global Wealth Report. Zurich: Credit Suisse Research Institute.

Credit Suisse (2011) Emerging Consumer Survey. Zurich: Credit Suisse Research Institute.

D'Andrea, G., Stengel, A. and Goebel-Krstelj, A. (2004) *6 Truths About Emerging-Market Consumers*. www.strategy-business.com/article/04106?gko=cac3d [accessed on 24 May 2015].

Das, D. (2009) Globalisation and an Emerging Global Middle Class. *Economic Affairs*, 29, pp. 89–92.

Dossani, R. and Kenney, M. (2009) Service Provision for the Global Economy: the Evolving Indian Experience. *Review of Policy Research*, 26(1–2), pp. 77–104.

Dubois, B. and Duquesne, P. (1993) The Market for Luxury Goods: Income versus Culture. *European Journal of Marketing*, 27(1), pp. 35–44.

Dubois, B., Czellar, S. and Laurent, G. (2001) *Consumer Rapport to Luxury: Analyzing Complex and Ambivalent Attitudes*. Paris: HEC School of Management.

Dubois, B., Czellar, S. and Laurent, G. (2005) Consumer Segments Based on Attitudes Toward Luxury: Empirical Evidence from Twenty Countries. *Marketing Letters*, 16(2), pp. 115–28.

Ernst and Young (2013) *Hitting the Sweet Spot: The Growth of the Middle Class in Emerging Markets*. Ernst and Young: London.

Evalueserve (2007) *India's Knowledge Process Outsourcing (KPO) Sector: Origin, Current State, and Future Directions*. Gurgaon: Evalueserve.

Fernandes, L. (2000) Restructuring the New Middle Class in Liberalizing India. *Comparative Studies of South Asia, Africa and the Middle East*, 20(1–2), pp. 88–112.

Fernandes, L. and Heller, P. (2006) Hegemonic Aspirations: New Middle Class Politics and India's Democracy in Comparative Perspective. *Critical Asian Studies*, 38(4), pp. 495–522.

Forbes (2011) *How the Uncertain Economy is Changing the Definition of Luxury*. www.forbes.com/sites/bluecarreon/2011/11/15/how-the-uncertain-economy-is-changing-the-definition-of-luxury [accessed on 15 February 2014].

Friedman, J. (1995) *Consumption and Identity*. London: Harwood Academic Publishers.

Gereffi, G., Humphrey, J. and Sturgeon, T. (2005) The Governance of Global Value Chains. *Review of International Political Economy*, 12(1), pp. 78–104.

Giddens, A. (1991) *Modernity and Self-identity: Self and Society in the Late Modern Age*. Oxford: Blackwell Publishing.

Goldman Sachs (2008) *The Expanding Middle: The Expanding Global Middle Class and Falling Global Inequality*. Global Economics Paper No 170. New York: Goldman Sachs.

Han, Y.J., Nunes, J.C. and Dreze, X. (2010) Signaling Status with Luxury Goods: The Role of Brand Prominence. *Journal of Marketing*, 74(4), pp. 15–30.

Holt, D.B. (1995) How Consumers Consume: A Typology of Consumption Practices. *Journal of Consumer Research*, 22, pp. 1–16.

Hubacek, K. Guan, D. and Barua, A. (2007) Changing Lifestyles and Consumption Patterns in Developing Countries: A Scenario Analysis for China and India. *Futures*, 39(9), pp. 1084–96.

Hudders, L. (2012) Why the Devil Wears Prada: Consumers' Purchase Motives for Luxuries. *Journal of Brand Management*, 19, pp. 609–22.

Jones Lang LaSalle (2014) *Retail Realty in India: A Comparison and Contrast with the Emerging Cities of Asia*. Chicago: Jones Lang LaSalle.

Kharas, H. (2010) *The Emerging Middle Class in Developing Countries*. Paris: OECD Working Paper 285.

Kirk, J. and Wall, C. (2011) *Work and Identity – Historical and Cultural Contexts*. Basingstoke: Palgrave Macmillan.

Kleibert, J.M. (2015) *Expanding Global Production Networks, The Emergence, Evolution and the Developmental Impact of the Offshore Service Sector in the Philippines*. Amsterdam: Dissertation Universiteit van Amsterdam.

Koefoed, L. and Simonsen, K. (2012) (Re)scaling Identities: Embodied Others and Alternative Spaces of Identities. *Ethnicities*, 12(5), pp. 623–42.

Krishnan, S. (2015) How the BPO Industry Contributes to the Formation of a Consumerist New Middle Class in Mumbai. In: Lambregts, B., Beerepoot N. and Kloosterman, R.C. (eds) (2015) *The Local Impact of Globalization in South and Southeast Asia: Offshore Business Processes in Services Industry*. London: Routledge, pp. 183–194.

Lambregts, B., Beerepoot, N. and Kloosterman, R.C. (eds) (2015) *The Local Impact of Globalization in South and Southeast Asia: Offshore Business Processes in Services Industries*. London: Routledge.

McKinsey (2007) *The 'Bird of Gold': The Rise of India's Consumer Market.* San Francisco: McKinsey Global Institute.

McKinsey (2013a) *Mapping China's Middle Class.* www.mckinsey.com/insights/consumer_and_retail/mapping_chinas_middle_class [accessed on 30 April 2015].

McKinsey (2013b) *Africa's Growing Giant: Nigeria's New Retail Economy.* www.mckinseyonmarketingandsales.com/africas-growing-giant-nigerias-new-retail-economy [accessed on 30 April 2015].

McKinsey (2013c) *Mapping the Mindset Of Brazil's Not-So-New Middle Class Consumers.* www.mckinseyonmarketingandsales.com/mapping-the-mindset-of-brazils-not-so-new-middle-class-consumers [accessed on 30 April 2015].

Murphy, J. (2011) Indian Call Centre Workers: Vanguard of a Global Middle Class? *Work, Employment and Society*, 25(3), pp. 417–433.

Nadeem, S. (2009) Macaulay's (Cyber) Children: The Cultural Politics of Outsourcing in India. *Cultural Sociology*, 3, pp. 102–22.

NASSCOM (2010) Impact of the IT-BPO Industry in India: A Decade in Review. New Delhi: NASSCOM.

Nijman, J. (2006) Mumbai's Mysterious Middle Class. *International Journal of Urban and Regional Research*, 30(4), pp. 758–75.

Nueno, J.A.L. (1998) The Mass Marketing of Luxury. *Business Horizons*, 41(6), pp. 61–68.

Pal, M. and Buzzanell, P. (2008) The Indian Call Center Experience. A Case Study in Changing Discourses of Identity, Identification, and Career in a Global Context, *Journal of Business Communication*, 45(1), pp. 31–60.

Price Waterhouse Coopers (2015) *Beyond the BRICS: How to Succeed in Emerging Markets (by Really Trying).* www.pwc.com/us/en/view/issue-15/succeed-emerging-markets.html [accessed on 24 May 2015].

Rastogi, G. and Pradhan, B. (2011) *Offshore: How India Got Back on the Global Business Map*. London: Penguin Books.

Times of India, The (2013) *Luxury Retail Space in India to Rise to 1.44% by 2015: Cushman & Wakefield.* http://articles.timesofindia.indiatimes.com/2013-09-26/indiabusiness/42425785_1_luxury-brands-mall-culture-mall-space [accessed on 7 January 2014].

Trentmann, F. (2010) Tiny Middle Class, Huge Moral Anxieties. *Economic and Political Weekly*, 29, pp. 35–36.

Upadhya, C. (2008) Rewriting the Code: Software Professionals and the Reconstitution of Indian Middle Class Identity. In: C. Jaffrelot and P. van der Veer (eds) *Patterns of Middle Class Consumption in India and China*. New Delhi: Sage: pp. 55–87.

Veblen, T. (1899) *The Theory of the Leisure Class: An Economic Study of Institutions.* MacMillan, New York.

Vigneron, F. and Johnson, L.W. (2004) Measuring Perceptions of Brand Luxury, *Journal of Brand Management,* 11(6), pp. 484–506.

Warde, A. (1994) Consumption, Identity-Formation and Uncertainty. *Sociology*, 28, pp. 877–98.

Warhurst, C. and Nickson, D. (2007) A New Labour Aristocracy? Aesthetic Labour and Routine Interactive Services. *Work, Employment and Society*, 21(4), pp. 785–98.

Wilson, D. and Dragusanu, R. (2008) *The Expanding Middle: The Exploding World*

Middle Class and Falling Global Inequality. Global Economics Paper 170: New York: Goldman Sachs.

UNCTAD (2012) *Information Economy Report 2012: The Software Industry and Developing Countries*. New York: United Nations.

Yong Jin, D. (2011) *Global Media Convergence and Cultural Transformation: Emerging Social Patterns and Characteristics*. Hershey: Information Science Reference.

Yoon, E. (2011) Measuring Ethnic Identity in the Ethnic Identity Scale and the Multigroup Ethnic Identity Measure-Revised. *Cultural Diversity and Ethnic Minority Psychology, American Psychological Association*, 17(2), pp. 144–55.

13 Conclusions

Trends and ways forward in services-driven economic growth

Niels Beerepoot, Jana Kleibert and Bart Lambregts

This book has highlighted a number of recent dynamics in the service sector. It started by arguing how services have transformed from a residual category of economic output to a dynamic sector that is the world's largest generator of added value and jobs. Case studies from different parts of the world were subsequently presented to underline how services play a role in providing new employment opportunities, are the most significant source of inward investment for many countries, and have become increasingly tradeable over a long distance. This final chapter brings together and reflects upon the theoretical and empirical insights from the case studies. The three tenets by Ghani and O'Connell (2014) that were introduced in Chapter 1 are used to structure the discussion in the first part of this final chapter.

The first tenet reads that services have long been thought to be driven by domestic demand and as such cannot by themselves drive growth. With the expanding tradability of services in the past few decades there is already plenty of evidence that this tenet or conventional wisdom no longer holds water. Since 1990, the share of services in world FDI stock has risen from 49 per cent to 63 per cent while the share of manufacturing has decreased (from 41 per cent to 26 per cent) and that of the primary sector has remained stable (at about 7 per cent) (UNCTAD, 2015). Different chapters in this volume have illustrated how the increasing global tradability of services continues to alter the global geography of services production and affects various dimensions of economic life in both the Global North and the Global South. In Chapter 2 it was demonstrated how business services are increasingly sourced globally with Asia and Central Europe emerging as new production locations. New destinations for offshore service delivery further include South Africa, which tries to obtain a share of the offshore service market (Chapter 10), and Northern Ireland, which recently has been successful in attracting inward investments in business services (Chapter 8). Cities such as Manila and Mumbai meanwhile have established themselves as important production centres in the ICT-ITES sector (Chapter 11 and 12). Tradeable services have led to impressive job gains in these locations, and have also stimulated local consumer-oriented services due to greater spending power of ICT-ITES workers. The emergence of new production locations can largely be attributed to the presence in many developing countries of an increasingly college educated labour force with sufficient English language skills. These destinations still confirm the 'traditional' North-South or core-periphery division in which work is relocated from high-cost environments to either peripheral areas in Western countries or to lower-wage countries. The same issue is also raised in Chapter 4, which showed how the most popular global online services (such as Facebook) originate from the Global North. Despite technology, capital and know-how being available, online service firms from the Global South barely manage to successfully market their products to an international audience. In Chapter 9 it was shown how emerging economies (BRICS and some East Asian countries) are slowly becoming more important as exporters of cultural industries products. Cultural industries are a good example of the blurring division between the industry and service sectors (for example architecture, game development), which makes singling out the role of services in driving economic growth or classifying certain activities as 'industrial' or 'service' more difficult.

In line with the cited UNCTAD figures, Chapter 6 demonstrated how FDI in services is the main form of inward investment in Great Britain (accounting for 68 per cent of incoming FDI in 2004–2005). At the national level, service investments disproportionately land in the more urban and the more peripheral areas with four service domains (Financial Services, Software and Computer Services, R&D and Business Services) accounting for the greatest share of the investments. Similar examples of the importance of services for economic development and economic resilience were found in Chapter 5, which demonstrated how in Europe employment growth largely takes place in knowledge-based and in leisure-based services. This underlines the need to further our understanding of which particular services have greater capacity to drive economic growth. Even within the producer-oriented services sector variety in this respect is large. In the Netherlands, for instance, employment agencies and the facilities management sector have been the first to benefit from the recent economic recovery (see CPB 2015). Yet, few would regard these sectors to be actual drivers of economic growth.

Chapters in this volume also demonstrated how increasing global service trade impacts on the local consumer service sector. In Chapter 5 it was argued how consumer-oriented services are an important component of a modern regional economy particularly through the inclusion of leisure and cultural services. In Amsterdam, the growth of international tourism has given a boost to the local restaurant sector (Chapter 7). This is accompanied by an increasing differentiation between the lower quality tourism market segment and the higher quality segment which mainly caters to better informed local clients. In Mumbai (as shown in Chapter 12) the growth of the ICT-ITES sector has led to quality upgrading in the local retail sector due to more spending power and more sophisticated demand from high earning young professionals. In sum, the chapters in this volume demonstrate that demand for services is increasingly global and the examples from (particularly) India and the Philippines show how tradeable, knowledge-based, services can stimulate economic growth. However, with the number of people involved in these activities in many countries being only a small part of the total services workforce, prudence is called for in gauging the potential scope of this effect.

The second tenet holds that services were long considered to have lower productivity than manufactures. While the work presented in this volume does not allow for a direct comparison of productivity levels in the services and manufacturing sectors, also since measuring productivity in many services remains a challenge, a number of observations can be made. In Chapter 3 it was demonstrated how a dynamic business sector is dependent on specialisation in high-tech export oriented manufacturing. This finding links well to the various publications that highlight the continued importance of the manufacturing sector as driver of economic development and innovation (e.g. Gereffi, 2014; Rodrik, 2015). In some (media) reports even a revival of manufacturing in Europe and the United States is observed (see *The Economist*, 2013). Automation, rising labour costs in the emerging economies of Asia, and concerns about having to rely on faraway places for the delivery of manufactured goods result in the return of factory production to Europe, but particularly back to the United States (e.g. Bailey and De Propris, 2014). While it is a welcome development for those seeking work in Europe and the United States, it will by no means compensate for the millions of manufacturing jobs that have been lost since the late 1970s. Therefore the service sector is, and likely will remain, the most important economic sector and source of FDI for most countries. The service sector is also where the biggest gains in productivity, for example through greater efficiency and adoption of ICTs, are still to be achieved (see Dable-Norris *et. al.*, 2015). So far, the relocation of service activities to South Africa, Northern Ireland, the Philippines or India, as case studies in this book showed, can be regarded as the 'low road strategy' in which achieving cost savings through labour arbitrage is considered a more feasible way to go than attempting to increase worker productivity 'at home'. The relocation of tradeable service activities has striking parallels with how the industrial sector, then especially enabled by lower freight transport cost, sought lower-cost production locations (see Fröbel *et al.*, 1980). Automation in industrial production eventually made labour cost relatively less of a concern (compared to other cost categories) enabling the reshoring of production as described above. When further advances in the application of ICT in services production (which initially enabled relocation to lower-cost locations) are going to allow for greater automation of the same tasks, services reshoring may become a reality as well, particularly when labour costs continue to rise in the Global South. A deeper understanding is therefore needed of the differentiation between service segments in terms of where greater opportunities for automation exist.

The third tenet is that it was long thought that 'good' jobs were created only in the manufacturing sector. This particularly refers to the historic role of the manufacturing sector as provider of employment to low skilled labourers (see Kloosterman *et al.*, 2015). Tradeable services are typically skill-intensive, and do not have the capacity to absorb – as manufacturing did – the type of labour that low- and middle-income countries have in abundance (Rodrik, 2015). Tradeable services mainly employ better educated workers in developing countries (as confirmed by Chapters 10 and 11). Recent debates on the 'servile class' or 'service underclass' (Scott, 2012; Standing, 2011) correctly point out that the 'cognitive-cultural' (knowledge) workers employed in advanced business services benefit from the low-end support services (for example security, housekeeping, transport) provided by the 'service underclass'. In recent years most new jobs in Western countries have been created in low-end services which are difficult to automate because they rely heavily on dexterity, flexible interpersonal communication, and direct physical proximity (Autor, 2013). Strong competition for these jobs will keep compensation low, leading to new concerns about polarisation in labour markets, both in Western countries (see Autor, 2013) and in the Global South (e.g. Krishna and Pieterse, 2008).

Call centres and back offices that relocated to India or the Philippines have also been critically scrutinised and are sometimes labelled 'digital sweatshops' (Taylor and Bain, 2005; Thite and Russel, 2010). The empirical examples from South Africa (Chapter 10) and the Philippines (Chapter 11) demonstrate that no sector other than IT-enabled services is currently able to provide developing countries medium skilled jobs in large numbers in only a short period of time. This important role in providing employment to an increasingly college educated labour force should not be underestimated. Earlier criticism on the quality of employment in call centres and back offices is hereby countered by a greater focus on the knowledge and skills gained during employment and opportunities for gaining first work experience. Chapter 11 demonstrated how work in the BPO sector stimulates the acquisition of generic skills which workers should be able to utilise in their future employment outside the BPO sector. Chapter 10 highlighted the attractiveness of the Legal Process Outsourcing sector for recent law graduates, illustrating the significance of the sector for gaining first work experience. More spending power of the workers in these activities subsequently leads to more sophisticated demand for (luxury) consumer goods and quality upgrading of the local retail sector as demonstrated in Chapter 12. Research elsewhere (see Beerepoot and Kumar, 2015) also showed how more advanced (tradeable) services lead to higher quality standards for low-end support services. The debate on the quality of employment in offshore services will continue, with most immediate threats for workers likely coming from labour saving automation (see Brynjolfsson *et al.*, 2014).

Perhaps the most untenable aspect of the three tenets, however, is that they take the service sector as a single, homogeneous category. The chapters in this book have dealt with different categories of services and shown how basically each of them is highly dynamic and capable of adopting different roles both in space and in time. Viewing services in the aggregate obscures this variety and makes it harder to see the development potential held by specific services activities. Opening up the heterogeneous and dynamic nature of services production is essential for enlarging our understanding of which particular services in which spatiotemporal context have the capacity to drive economic growth, have high(er) productivity, or provide 'good jobs'.

Avenues for further research

The work presented in this book has shed further light on how contemporary dynamics in services production generate new interdependencies between countries and regions across the globe, how they play a role in economic crisis mitigation and recovery in the Global North, and how they foster economic development and service sector-driven modernisation processes in the Global South. As such, and as shown above, the book has helped to further qualify a number of long adhered to tenets about the possible role of services in economic development. Yet, important issues and hence a great number of opportunities for future research on services remain. Perhaps the most critical question of all is whether the services sector can see sustained and strong enough productivity growth – and meanwhile continues to generate jobs – so as to fuel sustainable, service-led economic growth both in the Global North and in the Global South. At least four areas of inquiry can be identified where further research could contribute to answering this question.

First, technological advances have been driving the services transformation and their further developments and their diffusion will continue to change the limits of what role services can possibly play. While many commentators see increased automation as an inhibiting factor for employment creation, since fewer workers are needed, simultaneously examples of increased use for human service-labour arise out of precisely these new technologies that require human intelligence to be administered. One example are social media platforms and search engine optimisation, which are required by many of today's firms for the marketing of their services and products. These hybrid services require judgement and human intelligence and are ICT-enabled (Zysman *et al.*, 2013). Crowdsourcing, microsourcing and the 'human cloud', which involve a pool of virtual workers that conduct individual tasks for buyers, facilitated by mediating platforms, are new forms of digital service work that have emerged in recent years. Their reach is still limited to lower-end or more routinised tasks, and hampered by a lack of coordination and collaboration, as well as quality uncertainties (Kaganer *et al.*, 2013). However, it is a burgeoning phenomenon with potentially great impacts on the organisation of work, the inclusion of yet untapped human resources in the global economy, and thus on global wage development (see also Beerepoot and Lambregts, 2015; World Bank, 2015), and therefore deserves intense scholarly attention.

In close relation, the increasingly blurred line between industry and services is another phenomenon that requires further analysis (see also Lanz and Maurer, 2015). This blurring has led many traditional manufacturing firms (such as IBM, Philips, Apple and Volkswagen) to generate most of their profits from services. Several firms have shifted their core business from selling manufactured products to deliver services, which is helped by the fact that some products themselves can be turned into services (e.g. leasing a mobile phone becomes a service rather than the sale of a product, pay by use software on the internet, certain products are only useful as portals to access services) (Zysman *et al.*, 2013). These recent examples are linked to changing business strategies and models in which competition is increasingly taking place in the realm of service differentiation rather than product differentiation. Measuring the role of services in this is difficult, as is collecting general statistics on the sources and geographies of value creation in complex production networks.

Third, the geography of service delivery will continue to change and processes of service-sector decentralisation and (re)centralisation, of offshoring and reshoring, can be further empirically analysed in a number of sectors within the broad category of services in order to map these changes and better understand the underlying dynamics. The changing international and regional divisions of labour and the concomitant geographies of value creation and value capturing emerging are more complex than North-South patterns of trade and their drivers and inhibitors need to be scrutinised. Especially the emergence of lead firms in services originating from the Global South – to date most discussed are Indian ICT-ITES firms – may play a more crucial role in the future. The role of domestic-owned firms, rather than multinational corporation-owned subsidiaries, and their potential to upgrade and engender transformative growth in developing countries should be studied. What upgrading potentials exist for outsourced service operations and what limits are set by the organisational format of having several, often geographically dispersed and competing firms, delivering pieces of service work?

Finally, studies on the quality of service-sector employment can be envisioned. Especially the new forms of delivery models and networks, such as the 'human cloud' or 'platform economy' have implications for employment conditions around the globe. Is the breaking down of tasks into ever smaller components and singular transactions leading to a de-skilling of the labour force or are they democratising opportunities for income generation independent of location? Selling chunks of time to individual buyers via online platforms clearly increases employment insecurity. Yet, on the other hand, it allows for more decision-making freedoms on what offers to take on and at which prices by individual contractors, who do not have to sell their entire labour power to a single employer at a flat-rate. Until now, discussions in the Global North have focused mostly on the precariousness of the new work regimes created by the digital economy, whereas more hopeful visions prevail in countries in the Global South that are linking up to the digital economy. Integrating the analysis of winners and losers across geographies of these new service delivery patterns will remain an important task for and focus of scholarship in the years to come.

References

Autor, D.H. (2013). The "Task Approach" to Labor Markets: An Overview. Cambridge: National Bureau of Economic Research Working Paper 18711. Available at: www.nber.org/papers/w18711.pdf [accessed 1 November 2015].

Bailey, D. and De Propris, L. (2014). Manufacturing reshoring and its limits: the UK automotive case. *Cambridge Journal of Regions, Economy and Society*, 7(3), pp. 379–95.

Beerepoot, N. and Kumar, R. (2015). Upgrading service delivery and employment conditions through indirect insertion in global value chains. *Competition & Change*, 19(5), pp. 374–89.

Beerepoot, N. and Lambregts, B. (2015). Competition in online job marketplaces: towards a global labour market for service outsourcing? *Global Networks*, 15(2), pp. 236–55.

Brynjolfsson, E., McAfee, A. and Spence, M. (2014). New World Order: Labour, Capital, and Ideas in the Power Law Economy, *Foreign Affairs*, 93(4), Available online: www.foreignaffairs.com/articles/united-states/2014-06-04/new-world-order [accessed 3 November 2015].

CBS (2015). *Kwartaalmonitor zakelijke dienstverlening, tweede kwartaal 2015.* Centraal Bureau voor de Statistiek [Online] Available at: www.cbs.nl/NR/rdonlyres/ 77E9297D-13F3-4691-A920-8512A55B5C3F/0/monitorzakelijkedienstverlening2015kwartaal2.pdf [accessed on 4 November 2015].

Dabla-Norris, E., Guo, S., Haksar, V., Kim, M., Kochhar, K., Wiseman, K. and Zdzienicka, A. (2015). The New Normal: A Sector-Level Perspective on Growth and Productivity Trends in Advanced Economies. *IMF Staff Discussion Note.* Washington DC: International Monetary Fund. Available at: www.imf.org/external/pubs/ft/sdn/2015/sdn1503.pdf [accessed 16 November 2015].

Fröbel, F., Heinrichs, J. and Kreye, O. (1980). *The New International Division of Labour.* Cambridge: Cambridge University Press.

Gereffi, G. (2014). Global value chains in a post-Washington Consensus world, *Review of International Political Economy*, 21(1), pp. 9–37.

Ghani, E. and O'Connell, S.D. (2014). Can Service Be a Growth Escalator in Low Income Countries? World Bank Policy Research Working Paper 6971, Washington, DC: World Bank.

Kaganer, E., Carmel, E., Hirschheim, R. and Olsen, T. (2013). Managing the Human Cloud. *MIT Sloan Management Review*, 54(2), pp. 23–32.

Kloosterman, R.C., Beerepoot, N. and Lambregts, B. (2015). Service sector-driven economic development from a historical perspective. In: Lambregts, B., Beerepoot, N. and Kloosterman, R.C. (eds) *The Local Impact of Globalization in South and Southeast Asia: Offshore Business Processes in Services Industries.* London and New York: Routledge, pp. 17–28.

Krishna, A. and Pieterse, J.N. (2008). Hierarchical integration: The dollar economy and the rupee economy. *Development and Change*, 39(2), pp. 219–37.

Lanz, R. and Maurer, A. (2015). Services and global value chains – some evidence on servicification of manufacturing and service networks. WTO Working Paper ERSD-2015-03.

Rodrik, D. (2015). Premature Industrialisation. NBER Working Paper Series. Available at: www.nber.org/papers/w20935 [accessed 10 November 2015].

Scott, A.J. (2012). *A World in Emergence: Cities and Regions in the 21st Century.* Cheltenham: Edward Elgar Publishing.

Standing, G. (2011). *The Precariat: The New Dangerous Class.* New York: Bloomsbury Academic.

Taylor, P. and Bain, P. (2005). 'India calling to the far away towns': the call centre labour process and globalization. *Work, Employment and Society*, 19(2), pp. 261–82.

The Economist (2013). *Reshoring Manufacturing: Coming Home* [Online] Available at: www.economist.com/news/special-report/21569570-growing-number-american-companies-are-moving-their-manufacturing-back-united [accessed on 29 November 2015].

Thite, M. and Russell, B. (2010). The next available agent: work organisation in Indian call centres. *New Technology, Work and Employment*, 25(1), pp. 2–18.

World Bank (2015). The Global Opportunity in Online Outsourcing. Washington, DC: World Bank.

World Trade Organization, Economic Research and Statistics Division. Available at: www.wto.org/english/res_e/reser_e/ersd201503_e.pdf (accessed on 22 November 2015).

Zysman, J., Feldman, S., Kushida, K.E., Murray, J. and Nielsen, N.C. (2013). Services with everything: the ICT-enabled digital transformation of services. In: Breznitz, D. and Zysman, J. *The Third Globalization*, pp. 99–129. Oxford: Oxford University Press.

Appendix

Table A.1 Data availability by series in GGDC-10 sector database and Africa sector database

Country	*Acronym*	*Value added at current prices*	*Value added at constant prices*	*Employment*
Latin America				
Argentina	ARG	1950–2005	1950–2005	1950–2005
Bolivia	BOL	1958–2005	1950–2005	1950–2003
Brazil	BRA	1990–2005	1950–2005	1950–2005
Chile	CHL	1950–2005	1950–2005	1950–2005
Colombia	COL	1950–2005	1950–2005	1950–2005
Costa Rica	CRI	1950–2005	1950–2005	1950–2005
Mexico	MEX	1950–2005	1950–2005	1950–2005
Peru	PER	1991–2005	1950–2005	1960–2005
Venezuela	VEN	1960–2003	1950–2005	1950–2005
Asia				
Hong Kong	HGK	1970–2004	1974–2005	1974–2005
Indonesia	IDN	1966–2005	1960–2005	1971–2005
India	IND	1950–2005	1950–2005	1960–2004
Korea	KOR	1953–2005	1953–2005	1963–2005
Malaysia	MYS	1970–2005	1970–2005	1975–2005
Philippines	PHL	1971–2005	1971–2005	1971–2005
Singapore	SGP	1970–2005	1960–2005	1970–2005
Thailand	THA	1951–2005	1951–2005	1960–2005
Taiwan	TWN	1951–2005	1961–2005	1963–2005
Africa				
Botswana	BWA	1964–2010	1968–2010	1964–2010
Ethiopia	ETH	1961–2010	1961–2010	1961–2010
Ghana	GHA	1960–2010	1960–2010	1961–2010
Kenya	KEN	1960–2010	1964–2010	1969–2010
Malawi	MWL	1960–2010	1966–2010	1966–2010
Mauritius	MUS	1960–2010	1970–2010	1970–2010
Nigeria	NGA	1960–2010	1960–2010	1960–2010
Senegal	SEN	1960–2010	1970–2010	1970–2010
South Africa	ZAF	1960–2010	1960–2010	1960–2010
Tanzania	TZA	1960–2010	1960–2010	1960–2010
Zambia	ZMB	1960–2010	1965–2010	1965–2010

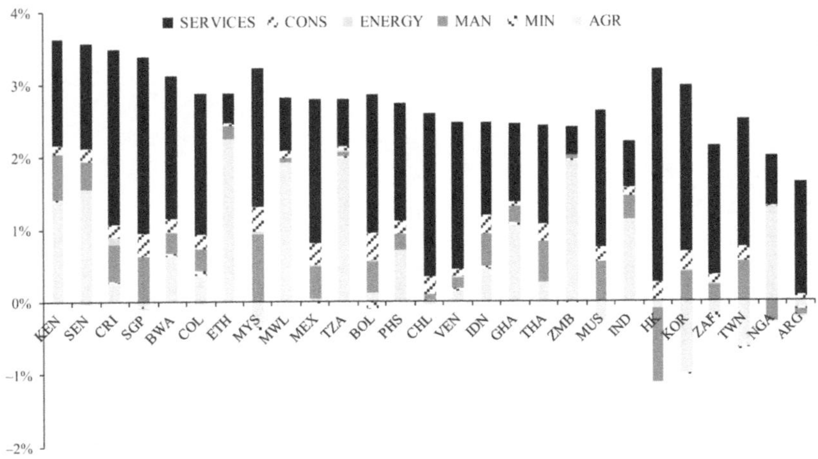

Figure A.1 Sectoral contributions to employment growth, by countries, 1975–2005

Notes: AGR: agriculture (sectors A+B of ISIC Rev.3.1), MIN: mining and quarrying (sector C), MAN: manufacturing (sector D), ENERGY: public utilities (sector E), CONS: construction (sector F), SERVICES: sectors G to Q of ISIC Rev.3.1. Bolivia and Venezuela 2003 instead of 2005; India 2004.

Source: Based on GGDC 10-sector database and the Africa sector database.

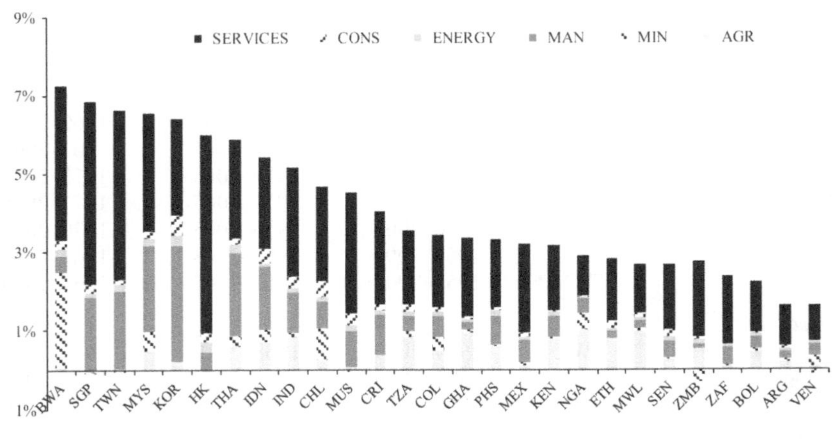

Figure A.2 Sectoral contributions to value-added growth, by countries, 1975–2005

Notes: Unweighted regional averages. AGR: agriculture (sectors A+B of ISIC Rev.3.1), MIN: mining and quarrying (sector C), MAN: manufacturing (sector D), ENERGY: public utilities (sector E), CONS: construction (sector F), SERVICES: sectors G to Q of ISIC Rev.3.1. Latin America (LATINAM) excludes Brasil and Peru due to data availability. Bolivia and Venezuela 2003 instead of 2005; India 2004. See Appendix for more details.

Source: Based on GGDC 10–sector database and the Africa sector database.

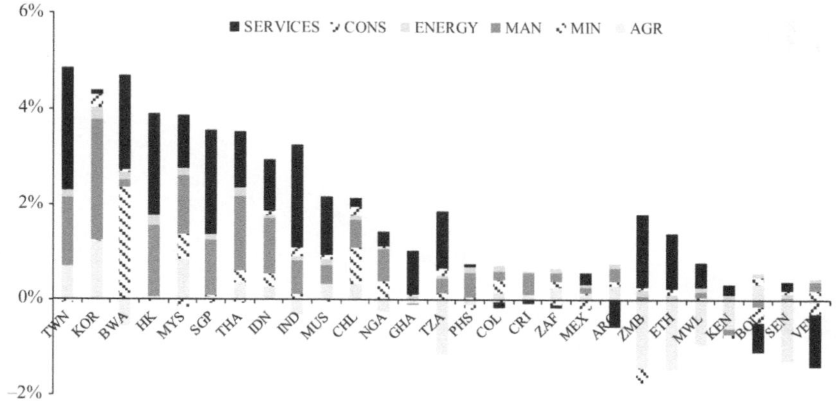

Figure A.3 Sectoral contributions to productivity growth, by countries, 1975–2005

Notes: Unweighted regional averages. AGR: agriculture (sectors A+B of ISIC Rev.3.1), MIN: mining and quarrying (sector C), MAN: manufacturing (sector D), ENERGY: public utilities (sector E), CONS: construction (sector F), SERVICES: sectors G to Q of ISIC Rev.3.1. Latin America (LATINAM) excludes Brasil and Peru due to data availability. Bolivia and Venezuela 2003 instead of 2005; India 2004. See Appendix for more details.

Source: Based on GGDC 10-sector database and the Africa sector database.

Index

For Product Safety Concerns and Information please contact our EU representative GPSR@taylorandfrancis.com Taylor & Francis Verlag GmbH, Kaufingerstraße 24, 80331 München, Germany